GitHub®

for **dummies**®
A Wiley Brand

GitHub®

2nd Edition

by Sarah Guthals, PhD

Contents at a Glance

Table of Contents

Introduction

Welcome to the world of collaborative coding! Whether you're just starting your coding journey, building fairly complex programs, or building with a team of people, this book guides you in using one of the most used tools for collaborative code-writing: GitHub.com. With more than 83 million users and over 130 million repositories (projects) hosted, GitHub.com is the No. 1 place to build and collaborate on code.

About This Book

Though you spend many hours sitting at your computer, alone, debugging and writing code, the ideal coding team includes more than just you. Hundreds of developers spent more than four years building World of Warcraft before its first release in 2004. Although occasionally you can build a big hit like Wordle alone in a couple of days, the norm for software development is that you will work with other coders, designers, testers, user experience experts, product managers, and sometimes hardware engineers to bring something to the hands of users.

When you're first starting out on complex coding projects, understanding effective ways to collaborate can be daunting. This book introduces you to the world of open source development (the epitome of collaboration), as well as effective ways to work with one other person — or even yourself over the course of many years! (I don't know about you, but Sarah from three years ago knows stuff that Sarah from today can't remember, and Sarah from today has more experience than Sarah from three years ago.)

GitHub For Dummies is written as a reference guide. Each part introduces you to a different aspect of collaborative coding. So if you're experienced in using GitHub, but you're new to the open source community, you can jump to Part 5 and skip some of the GitHub basics.

As you explore each part of this book, keep the following points in mind:

>> Words that are being defined appear in *italic*.

>> Code and URLs (web addresses) are shown in mono font.

>> Command sequences using onscreen menus use the command arrow. For example, when working in Scratch, you can open a new project as follows: From the menu bar, choose File ⇨ New.

>> The figures you see in this book use Mac and Chrome. I provide some tips when what you see on a Windows PC may be different, but you should see the same things, regardless of which Internet browser you use.

Foolish Assumptions

In this book, I make some assumptions that very well may be foolish, about you, your coding experience, and your goals.

>> You're interested in and have had some experience with coding. You don't have to be an expert coder, but you have made a Hello World application (or the equivalent) in at least one programming language.

>> You have patience and determination and are resourceful. When you're presented with a challenge, you can find a solution. This book guides you through GitHub.com as it exists at the time of writing it, but new features and workflows are being created, and part of your collaborative coding journey is to discover how to use those new features as they become available.

>> You have experience with a keyboard and mouse on either Mac or Windows PC and have access to one of those machines.

>> You're capable of using an Internet browser, such as Safari, Chrome, or Firefox, and you can type a URL to access a website, such as GitHub.com.

>> You know how to install applications on your computer. Although I guide you through anything that is unique to the setup, you should know how to download and install an application without step-by-step guidance.

Icons Used in This Book

Throughout the margin of this book are small images, known as *icons*. These icons mark important tidbits of information:

TIP

The Tip icon identifies places where I offer additional tips for making this journey more interesting or clear. Tips can start you on a rabbit hole down another work-flow, not covered in this book, or cover some neat shortcuts that you may not have known about.

REMEMBER

The Remember icon bookmarks important ideas to help you work more effectively throughout this book.

WARNING

The Warning icon helps protect you from common errors and may even give you tips to undo your mistakes.

Beyond the Book

In addition to what you're reading right now, this product also comes with a free access-anywhere Cheat Sheet that covers common commands and GitHub actions. To get this Cheat Sheet, simply go to www.dummies.com and search for **GitHub For Dummies Cheat Sheet**.

GitHub also offers Skills, which are free, guided tutorials that can be installed and found at https://skills.github.com/.

Where to Go from Here

GitHub is a tool used by millions of developers. The workflows that you discover in this book are just the beginning. As you become a more experienced coder, begin to collaborate on more elaborate projects, or join different companies and teams, you may encounter new workflows that use these tools in different ways. You should feel empowered to explore! Visit https://help.github.com or https://guides.github.com for guidance and don't forget to follow the blog at https://blog.github.com/ to stay up to date with all the new features!

1

Getting Started with GitHub.com

Discover how to use Git on your local computer to track changes in your project.

Sign up for a free GitHub.com account.

Explore GitHub.com resources and features.

Install GitHub Desktop to manage the link between your local and remote projects.

Install the Visual Studio Code editor as a lightweight option for coding.

Prepare for creating your own projects and contributing to others.

Chapter **1**

Understanding the Git in GitHub

Whether you're an experienced coder or a newbie starting out, learning how to work with others on code is critical to succeeding in the software industry. Millions of people around the world work together to build software, and GitHub is one of the largest tools to support a collaborative workflow. This chapter introduces you to the core tools you need to write code with other people.

Introducing GitHub

GitHub creates an environment that allows you to store your code on a remote server, gives you the ability to share your code with other people, and makes it easy for more than one person to add, modify, or delete code to the same file and project, while keeping one source of truth for that file (phew!). So what does that all actually mean? One of my favorite ways of explaining GitHub.com to folks who are new to the tool is to compare it to Google Docs — a place online where you can write code with other people and not have to email different versions back and forth.

What makes GitHub work behind the scenes is Git.

Understanding Version Control

Version control systems (also known as source control management, or SCM) are software that keep track of each version of each file in a coding project, a timestamp for when that version was created, and the author of those changes.

TIP

Writing code is an iterative process. For example, when you're building a website, you first may want to get some basic structure up before adding all your content. The best thing to do is to create a version of your website each time you have something that works. That way, as you experiment with the next piece, if something breaks, you can just go back to your previous version and start over.

SCMs enable coders to make mistakes without worrying that they'll have to completely start over. Think of it like being able to click Undo, but instead of undoing each key press, you can undo an entire piece of the project if you decide you don't like it or it doesn't work.

The basic workflow of coding with version control system support is as follows:

1. **Create a project, typically in a folder on your computer.**

2. **Tell your version control system of choice to track the changes of your project/folder.**

3. **Each time your project is in a working state, or you're going to walk away from it, tell your version control system of choice to save it as the next version.**

4. **If you ever need to go back to a previous version, you can ask your version control system to revert to whichever previous version you need.**

You can use a version control system if you're working alone on your own computer, but it gets even more interesting when you begin working with other people. (For more on working with other people, see the section "Git Version Control," coming up next in this chapter).

For more information about version control, visit `https://git-scm.com/book/en/v2/Getting-Started-About-Version-Control`.

Git Version Control

GitHub, as the name would suggest, is built on Git. Git is a type of version control system, and it's free and open source, which means that anyone can use it, build on top of it, and even add to it.

GitHub products make using Git easy, but if you're curious, you can also use Git to track your solo projects on your computer. You can find a brief introduction to local Git commands for solo projects in the next section.

Try simple Git on the terminal

With the help of Git for Windows, using the terminal on Mac, Windows, or Linux computers is exactly the same. A *terminal* is an application that enables you to interact with your computer in a text-based way — in other words, instead of double-clicking and dragging, you type commands to navigate your computer.

If you're on Mac or Linux, a terminal is already installed on your computer. If you're using a Windows computer, you have a couple options:

>> You can use the Windows Terminal and install the Windows Subsystem for Linux (WSL) and you can follow the same instructions as a Mac or Linux terminal.

>> You can install Git for Windows. Just go to https://gitforwindows.org and click Download to gain access to Git Bash, an emulator that allows you to interact with Git just like you would on a Linux or Mac terminal. You also get Git GUI, which gives you a user interface for almost all Git commands you might type into Git Bash, and shell integration so that you can quickly open Git Bash or Git GUI from any folder.

Many developers on Windows prefer to use PowerShell as their terminal environment. You can use Git within PowerShell, but setting that up properly is outside the scope of this book. However, you can find a handy guide to setting this up at https://haacked.com/archive/2011/12/13/better-git-with-powershell.aspx.

The Windows Subsytem for Linux (WSL) lets developers run a GNU/Linux environment directly from Windows. You can learn more about it on the Microsoft docs page https://learn.microsoft.com/windows/wsl/.

First, find the Terminal application:

>> On Mac, click the magnifying glass at the top right of your desktop, type **Terminal**, select the terminal from the list of applications, and press Enter or click it.

>> On Linux, press Ctrl-Alt-T at the same time, and the terminal window opens.

» On Windows, click the Windows menu in the bottom right of your desktop, search **Windows Terminal** or **Git Bash,** select the application from the list of search results, and press Enter or click it.

When the application opens, type `git --version` in the terminal. If you have Git installed, you should see a version number, as shown in the following code (the $ is a common indicator that the terminal is ready for input and is often already on the line; when you see that throughout this book, you should not type it). Otherwise, you can follow the instructions on https://git-scm.com/book/en/v2/Getting-Started-Installing-Git.

WARNING

When using the command line, you have to be very careful about exactly what you're typing. In the following code, the first instruction is for you to type `git --version`. You should note that a space appears between `git` and the rest of the instruction but no other spaces. You should also note the two dashes before the word `version`. They can be easy to miss, so be careful!

For Mac or WSL/Linux, you should see something like this:

```
$ git --version
git version 2.37.2
$
```

For Windows, you should see something like this:

```
$ git --version
git version 2.37.2.windows.2
$
```

Next, using the terminal, go to your desktop and create a new folder called git-practice. To do this, you should type the following commands:

```
$ cd ~/Desktop
$ mkdir git-practice
$ cd git-practice
$
```

For Mac or WSL/Linux if you type `pwd`, you should see that you are now in the folder git-practice, which is on your desktop. It might look something like this:

```
$ pwd
$ /Users/drguthals/Desktop/git-practice
$
```

If you're using the command prompt in Windows instead of Git Bash or WSL, you should use cd to print the current directory instead of pwd.

TIP

In 2020, GitHub heard the developer community and began a massive renaming project to stop using "master" as the default name for the primary branch of a respository and to use "main" instead. This change has proliferated to Git as well. When you initialize your local folder to use Git, you might be prompted to update your default primary branch name to "main." Whether you are prompted or not, you can update all Git repositories to use "main" as the primary branch name with this command:

```
$ git config --global init.defaultBranch main
```

I recommend that you run this config command before you initialize your Git repository so that your primary branch is called main.

Now, you can tell Git to track this folder using the init command.

```
$ git init
Initialized empty Git repository in /Users/drguthals/Desktop/
   git-practice
$
```

Then make sure that you have a clean folder. You can check with the status command:

```
$ git status
On branch main
No commits yet
nothing to commit (create/copy files and use "git add" to track)
$
```

Then, you can create a file to have Git start tracking and confirm the file is in the folder:

```
$ echo "practicing git" > file.txt
$ ls
file.txt
$
```

On Mac, you can open this folder in a Finder window with the open <path> command:

```
$ open .
$
```

On Linux, you can open this folder with the `nautilus <path>` command:

```
$ nautilus .
$
```

On Windows, you can open this folder with the `explorer <path>` command:

```
$ explorer .
$
```

This puts `.` as the `<path>` for each command. The period (`.`) tells the terminal to open the current folder. You could also use a different path with these commands to open other folders.

After the folder is open, double-click the file called `file.txt`, and the file opens with TextEdit on Mac, gedit on Linux, and Notepad on Windows. You can see that the words "practicing git" are actually there.

Close the file. Now, you can tell Git that you want to save this as a particular version. Back in the terminal:

```
$ git add file.txt
$ git commit -m "Adding my file to this version"
[main (root-commit) 8d28a21] Adding my file to this version
1 file changed, 1 insertion(+)
Create mode 100644 file.txt
$ git status
On branch main
nothing to commit, working tree clean
$
```

You can make a change to your file in the text file. Open the file again, change the text to say "Hi! I'm practicing git today!" and then choose File ⇨ Save and close the text application.

When you go back to the Terminal to check the status of your project again, you should see that Git has noticed that the file has changed:

```
$ git status
On branch main
Changed not staged for commit:
(use "git add <file..." to update what will be committed)
(use "git checkout -- <file>..." to discard changed in working directory)
```

```
modified: file.txt
no changed added to commit (use "git add" and/or "git commit -a")
$
```

`Commit` this version of your file again and notice that Git recognizes that every-thing has been saved to a new version:

```
$ git add file.txt
$ git commit -m "I changed the text"
[main 6d80a2a] I changed the text
1 file changed, 1 insertion(+), 1 deletion(-)
$ git status
On branch main
nothing to commit, working tree clean
$
```

If your terminal starts to get too cluttered, you can type `clear` to clear some space and make it more visually appealing. Don't worry; you can always scroll up and see everything you typed earlier!

Say that you actually want to see the original change, when you added "practicing git". First, get the `log` of all the `commits` you have made:

```
$ git log
commit 6d80a2ab7382c4d308de74c25669f16d1407372d (HEAD -> main)
Author: drguthals <sarah@guthals.com>
Date: Sun Aug 7 08:54:11 2022 -0800
I changed the text
commit 8d28a21f71ec5657a2f5421e03faad307d9eec6f
Author: drguthals <sarah@guthals.com>
Date: Sun Aug 7 08:48:01 2022 -0800
Adding my file to this version
$
```

Then ask Git to show you the first `commit` you made (the bottom most one). Make sure that you're typing your unique `commit` hash. In this book, the hash starts with 8d28a2. Make sure you type the entire hash that appears in your Git log.

Instead of typing the entire hash (and possibly having a typo), you can highlight the hash with your mouse, right-click and choose Copy, and then after `git checkout`, you can right-click and choose Paste. Using the keyboard shortcuts Ctrl+C or ⌘-C doesn't work.

```
$ git show 8d28a21f71ec5657a2f5421e03faad307d9eec6f
commit 8d28a21f71ec6567a2f5421e03faad307d9eec6f
```

```
Author: drguthals <sarah@guthals.com>
Date: Sun Aug 7 08:48:01 2022 -0800
Adding my file to this version
diff --git a/file.txt b/file.txt
new file mode 100644
index 0000000..849a4c7
--- /dev/null
+++ b/file.txt
@@ -0,0 +1 @@
+practicing git
$
```

You can see that practicing git was added to the file in that original commit.

For more information on how to use Git on the command line, check out the following resources:

>> The GitHub Git Cheat Sheet at https://education.github.com/git-cheat-sheet-education.pdf

>> The Visual Git Cheat Sheet at http://ndpsoftware.com/git-cheatsheet.html

>> The Git Docs page at https://git-scm.com/doc

Another resource for learning and understanding Git is https://learngit branching.js.org. This is a good self-guided set of exercises.

Git branching by collaborator

Git is different from other version control systems because it has fast branching, shown in Figure 1-1. *Branching* is a Git function that essentially copies code (each branch is a copy of the code), allows you to make changes on a specific copy, and then merges your changes back into the primary (main) branch.

When you're writing code, you will add files and commit changes to your main branch. Figure 1-1 outlines a specific workflow where two people are collaborating on the same file. Person 1 creates a new branch called MyBranch and makes some changes to the file. Person 2 also creates a new branch called YourBranch and makes some changes to the same file. You can see this change in box #1.

You can see the difference (called diff) between the main branch and MyBranch in Figure 1-1 in box #2.

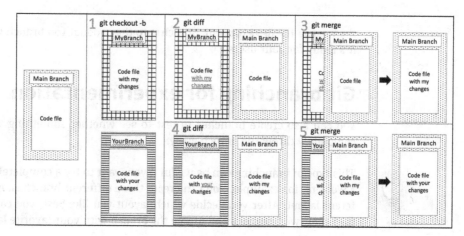

Then, Person 1 merges their changes with the main branch, as you can see in box #3.

Person 2 has made their own changes, but before merging, they will make sure they have the most updated version of the main branch, which now has the changes from Person 1. The diff can be seen in box #4. Notice what text is in both files.

Finally, Person 2 acknowledges that their changes will overwrite Person 1's changes and merges their changes with main branch, making the final version have the changes from Person 2. Box #5 shows this final merge, with the main branch having the final changes.

REMEMBER

Figure 1-1 shows just one workflow that can exist when more than one person is working on code and is meant to describe branching. You can get a more in-depth overview on Git and branching at https://git-scm.com.

Git branching by feature

Another common way to use branching is to have each feature that you develop be on a different branch, regardless of the collaborator building the feature.

You can extend the idea of branching by feature to also have one branch that is your production branch. This branch is what your users will see. Then you can have a development branch, which is one that you can merge features into without changing what your users see.

This type of branching allows you to build a lot of different features, merge them each into the development branch, make sure they all work the way you want, and then merge the development branch into the production branch when you know it's ready for your users.

Git branching for experimentation

You can also create branches to test to see whether something works and then completely throw the branch away.

TIP

This type of branching can be useful if you want to try a completely new layout of a website, for example. You can create three different branches, each with a different layout. After you decide which layout you like best, you can simply delete the other two branches and merge the branch with your favorite layout into main.

Git's Place on GitHub

GitHub is a host for Git repositories. At some point, it's helpful to place your Git repository in a shared location as both a backup and a place where others can collaborate with you on your code. As a Git host, GitHub provides all the features of Git in addition to a few extra useful services.

REMEMBER

On GitHub.com, projects, or *repositories*, are stored on remote GitHub servers. If you save all your code on GitHub.com and your computer crashes, you can still access it.

Here is a list of some core Git features that GitHub supports:

>> **Repository:** Each repository contains all the files and folders related to your project and gives you control of permissions and collaborators' interaction with your code.

>> **Clone:** When you want to make changes to your code, you will often want to create a copy, or clone, of the project on your local computer. The cloned project is still tightly connected with the version on GitHub.com; it's just your local copy.

>> **Fork:** *Forking* a project is when you create your own copy of the entire project. When you fork a project, GitHub.com creates a new repository with your copy of all the files. You can still suggest changes back to the original copy, but you can also take your version and go in a new direction.

>> **Branches:** GitHub.com supports branching and even provides a useful tool — *pull requests* — to compare the diff between branches and merge branches.

>> **Commits:** GitHub.com tracks all the commits that you push to its servers and gives you an easy interface for browsing the code at different branches and different commits.

Signing Up for GitHub.com

GitHub.com offers unlimited free public and private repositories for individuals. Free private accounts are limited to three collaborators. You can sign up for a paid account to have unlimited collaborators and some Pro features.

Public means that anyone can see your code, clone your code, and therefore use your code. GitHub is a huge supporter of *open source software* (OSS) and therefore encourages public, shared code. Open source software is more than just public, shared code (see Part 5). Because every line of code can be traced to an author, you still get credit for what you've written, but the goal is to keep the code available for anyone to use, extend, and explore.

The following steps guide you through signing up for a free, individual GitHub.com account:

1. **Go to GitHub.com and click Sign Up.**

2. **Complete the Sign Up form.**

   This form helps GitHub understand who is using the software and helps them support workflows specific to their users. It also helps them suggest the best plan for what you need.

3. **Choose the plan you want.**

   For the purpose of this book, you can use the Free plan. You can always upgrade to a paid plan later if you decide you want to have more than three collaborators for your private repository and other Pro GitHub features.

   You're now at the home page, shown in Figure 1-2.

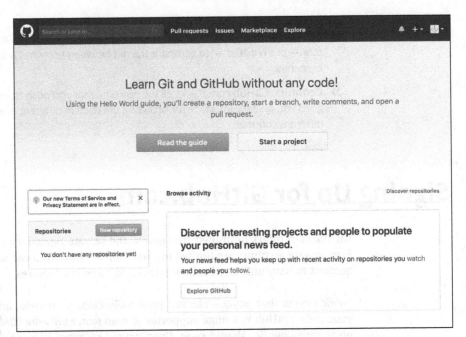

FIGURE 1-2:
The GitHub.com home page when you're logged in.

Personalizing Your GitHub.com Account

As you become a more experienced coder, you may want to reference your GitHub.com profile on your resume and job applications. More and more companies care more about your portfolio than a list of degrees or awards. For example, GitHub doesn't require you to provide information on your education as part of the hiring process and instead asks for a link to your GitHub.com profile and/or portfolio.

To complete your GitHub.com profile:

1. Click the avatar icon in the top-right corner and choose Settings.

2. Fill out the form on the Public Profile Settings page.

3. Click Update Profile when you're finished.

On the Personal Settings page, you can also adjust a number of different settings to continue personalizing your account.

Account

In Account settings, you can change your username, export your repositories and profile metadata, assign a successor to transfer your account to them in the event of your death, or delete your account.

WARNING

Changing your username may cause unintended side effects, so it typically isn't recommended. Just make sure that after you change your username that anything that you need to continue working still does. Follow links, test code, and run your applications again.

Appearance

Developers often have strong preferences between "light mode" and "dark mode." On the Appearance settings page, you can specify between them, and even specify secondary colors. If you want to make sure GitHub.com is synced with your system preferences, you can even do that. Additionally, you can specify the skin tone of the emojis you use, tab sizes, and whether you want to use monospace for editors within GitHub.com that support markdown (for example, Issue and Pull Request descriptions).

Accessibility

GitHub now supports a growing set of accessibility features including keyboard shortcuts and motion settings. At the time of publication of this version, motion settings are focused on whether gifs are autoplayed.

Notifications

Notifications can get really overwhelming. Though you can choose your level of granularity for receiving notifications per repository, this page displays your default preferences for notifications.

WARNING

I recommend not automatically watching repositories because any kind of activity that happens on any repository that you interact with will start to show up in your inbox, which turns out to not be helpful as you begin collaborating more.

TIP

Click the Things You're Watching link at the top of the notifications settings page to check to see what you're watching and therefore what notifications you may get from them.

Billing and plans

You can upgrade your plan at any time to include Pro features, such as unlimited collaborators and advanced code review tools. You can make this upgrade happen on the Billing settings page. In addition to upgrading your plan, you can also purchase add-ons such as GitHub Co-pilot, Git LFS data, and other Marketplace Apps.

This page also contains any open source projects you're sponsoring through the GitHub Sponsors program. You can learn more about GitHub Sponsors at https://github.com/sponsors.

TIP Git LFS stands for Git Large File Storage. Some software development requires large files, such as game scenes in video game development, to be stored. Without Git LFS, you can upload files as large as 100MB. Anything larger requires Git LFS, which supports files up to 2GB.

Emails

GitHub allows you to link multiple email addresses to your account. Notice that you can add email addresses, keep your email address private, and even block Git commands that may expose your email address.

Passwords and authentication

On this settings page you can update your password, add or update your two-factor authentication, and view a list of every computer address, city, and country where you're logged in or connecting to GitHub.com. *Two-factor authentication* means that when you type the correct password, you're asked to further verify that it is you who is attempting to log in through an app or SMS.

This page also includes a list of mobile devices where you have the GitHub app installed and you're logged in; this is a great way to quickly authenticate on GitHub.com on a new device.

SSH and GPG keys

At some point, you may want to create an SSH or GPG key to encrypt your communication with GitHub and ensure a secure environment. You can do this in your settings.

SSH keys enable you to connect to GitHub from your local machine without having to put in your username and password each time. GPG keys mark tags and commits that you make as verified, meaning that other people know that it was actually you who pushed the changes.

TIP Another way to tell Git to remember your credentials is to use a credential helper. GitHub tends to recommend this over using SSH, especially for Windows users. For more information on how to set up this feature, visit https://help.github.com/articles/caching-your-github-password-in-git.

Organizations

Organizations enable you to put GitHub users and repositories under similar settings. For example, you can grant admin rights to all repositories in an organization to the entire organization, instead of having to individually add each person to each repository. Although Organizations is out of the scope of this book, you can read about them on the GitHub Help page at https://help.github.com/articles/about-organizations.

Moderation

Moderation settings are split into three categories: users, repository interactions, and code reviews. These settings are specifically added to keep you safe from harassment and abuse. You learn more about building safe communities on GitHub from the Community section on the GitHub Docs, https://docs.github.com/communities.

Blocked users

In the Blocked users settings, you can block users from all your repositories.

Interaction limits

You can limit interactions on all of your repositories at once to only existing GitHub users, existing contributors on each repository, or existing collaborators on each repository. You can use this setting to force a "cooling off" period and is particularly useful if you're involved in heated discussions a GitHub user, and not just a specific repository. Interaction limit settings within your user settings override any existing interactive limits you may have on individual, public repositories.

Code review limits

You can also limit who can approve or request changes on pull requests across your repositories. Like with interaction limits, any code review limits set within your user settings will override existing code review limits you may have on individual, public repositories.

Repositories

The Repositories section lists all the repositories that you have created or been invited to as a collaborator. You also can leave repositories from this page. This section is where you can set your default name for primary branches on new repositories that you create. The default for GitHub is "main."

Packages

The Packages section contains any packages that you created and deleted. You have up to 30 days to restore a deleted package, and then it's deleted permanently.

GitHub Copilot

GitHub Copilot is a feature that was released in 2022 that uses artificial intelligence to suggest code (even entire functions) in real-time from within your editor. At the time of writing this version of the book, the cost for GitHub Copilot is $10 a month or $100 a year. You can learn more on the feature page at https://github.com/features/copilot/.

Pages

Chapter 5 walks you through creating a web page through the GitHub Pages feature. The Pages settings page is where all the verified domains that you own and have connected to GitHub are listed. Each of these can be used within your individual repositories.

Saved replies

Saved replies can be extremely useful for popular OSS. For example, if you're building an extension to an application, a lot of folks may report problems with the application, not with your extension. You can write a saved reply to point folks to where they can provide feedback on the application when they find an error.

Code security and analysis

One of the easiest ways to introduce security vulnerabilities into your applications is through libraries and packages that you depend on. In the Code security and analysis settings, you'll find the option to turn on dependency graphs, dependabot alerts, and dependabot security updates for each new repository.

Applications

You can connect three kinds of applications with your GitHub.com account:

>> **Installed GitHub apps:** GitHub applications that you are using with your account. One example is GitHub Learning Labs.

>> **Authorized GitHub apps:** Applications that you have authorized to access your account. One example is Slack.

>> **Authorized OAuth apps:** Applications that you have authenticated with using GitHub credentials. One example is GitHub Desktop.

Scheduled reminders

You can set up reminders to help prioritize your most important tasks first. Reminders will be sent to you via a Slack message, but require you to link your Slack workspace with your GitHub account.

Security log

This section in your settings lists recent events related to your GitHub account, such as sign-ins, with information such as the IP address of the computer where the action was performed, when the event happened, and a physical location of the computer. This is helpful for monitoring access to your account to ensure your security.

Sponsorship log

The Sponsorship log contains recent activity on your own sponsorships. You are notified here if you receive new sponsorships, someone changes their sponsorship, or someone cancels their sponsorship.

Developer settings

The last section on the settings page is Developer settings, which you use only if you're building an application that accesses the GitHub API, which means the application needs to access GitHub data in some way.

Three settings appear in this section:

>> **OAuth apps:** Applications you have registered to use the GitHub API.

>> **GitHub apps:** Applications that integrate with and extend GitHub.

>> **Personal access tokens:** Similar to SSH keys, tokens allow you to access the GitHub API without requiring authentication.

Discovering Helpful Resources

The GitHub.com help page (https://help.github.com) has an extensive list of documents for every feature on GitHub.com. From the top-right avatar menu, you can click the Help link. From there, clicking Contact a Human takes you to the GitHub contact page (https://support.github.com/), where you can find the following resources:

>> A FAQ

>> Links to the help documentation

>> Links to developer documentation for help on the GitHub API (https://docs.github.com/)

>> The GitHub Skills for guided GitHub exercises (https://skills.github.com/)

>> The GitHub Community Forum, where you can ask questions and get answers from folks who work at GitHub and other community members (https://github.community)

>> A slew of other resources about your experiences on GitHub.com

Chapter **2**

Setting Up Your Collaborative Coding Environment

G itHub has a lot of features to offer new and returning coders. A good way to get acquainted with all those features is to explore the GitHub.com website. This chapter not only gives you that overview, but also guides you in setting up your local machine so that you can start building.

Exploring GitHub.com

The home page of GitHub.com, shown in Figure 2-1, is a great starting point for many tasks, including starting your own project, learning about a topic, or exploring existing repositories.

The top menu bar, shown in Figure 2-2, is always available to you and is a direct link to the most important functions you need to perform.

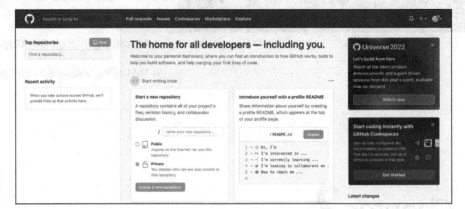

FIGURE 2-1:
The home page of GitHub.com.

Account menu

GitHub logo   Notifications

FIGURE 2-2:
The top menu bar of GitHub.com.

Search bar   Quick pick

» **GitHub home page:** If you click the GitHub logo in the top left of the browser, you return to the home page. Check out the sidebar "Mona Lisa Octocat" for more information on the logo.

» **Search bar:** The search bar on the top menu is pretty snazzy. Not only can you search all of GitHub, but as you start to use the site, it offers suggestions based on your most recent activity. These suggestions make it fast and easy to find the repository you were working on yesterday.

» **Pull requests:** The link to pull requests takes you to a list of all pull requests that you created, were assigned to complete, were mentioned in, or were asked to review. A pull request is a proposed change to the code of a repository. When first starting, you normally don't have anything in this section, but as you start interacting with collaborative repositories, you get an overview of any tasks you may want to attend to. For more on pull requests, see Chapter 3.

TIP

If you click the Pull Requests link, you might notice a ProTip, shown in Figure 2-3. The search bar for pull requests gives you several ways to specify a search to get exactly what you're looking for. In fact, an entire page (https://help.github.com/articles/searching-issues-and-pull-requests) is dedicated to effective searching. You can find ProTips throughout GitHub.com, so be sure to look out for them.

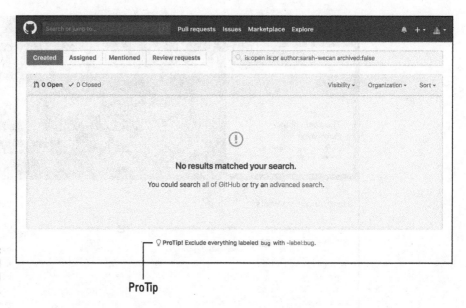

FIGURE 2-3:
ProTip found
on the pull
request page.

ProTip

>> **Issues:** The list of issues is almost the same as the list of pull requests. The main difference between an issue and a pull request is that an issue is a report of a bug or a feature request. An issue doesn't contain a proposed code change like a pull request does and therefore doesn't require a reviewer.

>> **Marketplace:** The marketplace on GitHub is a great place to find applications and tools that can help your collaborative coding workflow. For example, I have used AppVeyor, a continuous integration application, on projects. When you connect AppVeyor to one of your repositories, it continuously runs tests and deploys apps to make sure that every bit of code you're adding won't break what you've already built.

>> **Explore:** The Explore link takes you to a list of things you may be interested in (see Figure 2-4). You may find events and opportunities that GitHub hosts or supports. For example, GitHub released "The State of the Octoverse," which presents a lot of interesting analytics about code on GitHub — for example, 94 million developers on GitHub made 413 million contributions in 2022!

>> **Notifications:** The bell icon leads you to a list of your notifications. See Chapter 1 for how to change your notification settings.

>> **Quick pick:** The add-sign icon provides you with a list of quick actions you can take at any time: create a new repository (coding project), import a repository from another SCM, create a *gist* (a quick way to share code, notes, and snippets), or create a new organization.

>> **Account menu:** The account menu appears when you click your avatar. Here, you can get to your profile, repositories, anything you've starred, gists you've created, the help documents, settings, and sign out.

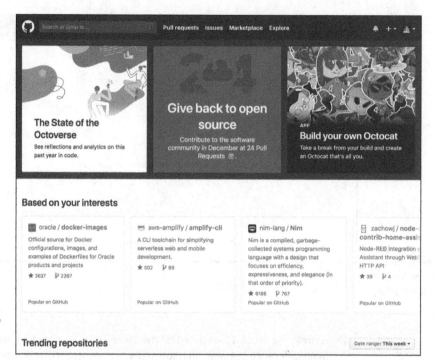

FIGURE 2-4:
Curated list of
repositories on
GitHub.com.

MONA LISA OCTOCAT

The GitHub logo is a 2D rendition of an *octocat* — a cat with five octopus-like arms. The logo was found on iStock, a website where royalty-free digital images can be purchased. Simon Oxley, the original designer, is also known as the designer for the Twitter bird logo. Cameron McEfee then led the effort around creating an entire Octodex of Octocats, which you can see at `https://octodex.github.com`.

One of the most popular things Mona has released is the "Build Your Own Octocat" app, which you can find at `https://myoctocat.com/build-your-octocat`. I created one and found my inner super woman!

To discover the full history of the Octocat, visit `http://cameronmcefee.com/work/the-octocat`.

Understanding Your Profile

Your profile is a public view of, essentially, your portfolio. To view your profile, click your avatar and then choose Your Profile from the menu that appears. You can see my profile in Figure 2-5.

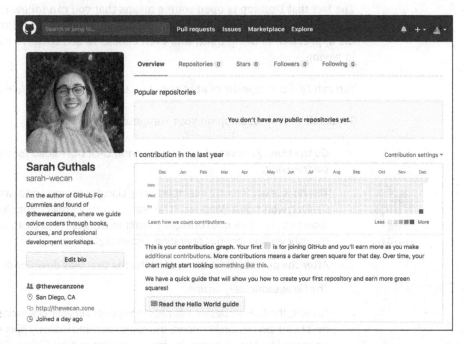

The top menu bar of your profile offers quick links to your repositories, things you've starred, your followers, and anyone who you follow. Below that are repositories that you're often visiting and your contribution graph. The *contribution graph* tracks how much code you've written per day. You can choose to include contributions to private repositories and an activity overview, which is a new feature.

WARNING

The amount and frequency at which you write code is not the bar by which software development is measured. It's true that the more you practice, the better you will be, but the practice you do must be deliberate. Making random code changes every single day without challenging yourself, or giving yourself time to think, design, and plan the code you want to write, is much worse than missing a white square.

Getting to Know GitHub Desktop

GitHub Desktop is a free, open source application that makes it easier for Mac and Windows users alike to manage repositories and GitHub connections on their local computer.

The fact that Desktop is open source means that you can follow the development of new features, connect with the developers right on the actual repository where the application is being built, and even choose to add features you're interested in having.

You can find the repository at `https://github.com/desktop/desktop`.

To install GitHub Desktop on your computer:

1. **Go to `https://desktop.github.com` and click Download for the platform you're using.**

REMEMBER

   This book is written with examples using Google Chrome and a Mac. GitHub Desktop works on a Windows PC as well because it's built using Electron, which allows it to work on both operating systems. Double-check that you download the right version for your operating system and browser.

2. **After the download finishes, click the file that was downloaded.**

   The file automatically unzips.

   On Mac, the GitHub Desktop application appears in your Downloads folder, next to the zip file. On Windows, the application immediately opens after you unzip the file. If you run into any issues, you can visit the GitHub Desktop Docs pages at `https://docs.github.com/desktop`.

3. **On Mac, drag the purple GitHub Desktop application into your Applications folder.**

4. **On Mac, go to your Applications folder and double-click the GitHub Desktop icon.**

   The application opens, shown in Figure 2-6.

WARNING

You may get an alert that you 're trying to open an application that was downloaded from the Internet. Click Open if this alert appears.

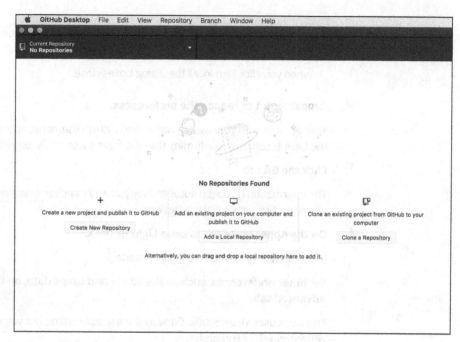

FIGURE 2-6:
The GitHub
Desktop
application
default view.

Setting Up GitHub Desktop

Before you can use GitHub Desktop effectively, you have to do a few things to connect it to your GitHub.com account. If you do not yet have a GitHub.com account, go to Chapter 1. If you already have a GitHub.com account and have already downloaded GitHub Desktop, you can set up GitHub Desktop with the following steps:

1. **Open the GitHub Desktop application.**

2. **Sign in to your GitHub or GitHub Enterprise account.**

   GitHub Enterprise accounts are typically used for companies that choose to host GitHub separate from the rest of GitHub.com.

3. **Specify your name and email that will be associated with your git commits.**

   If you decide to skip signing into GitHub now, you can always sign in later following these steps:

   1. Choose GitHub Desktop ⇨ Preferences.

   2. On the Accounts tab, click Sign In on the GitHub.com row.

      The Sign In dialog box, shown in Figure 2-7, appears.

3. Type your username and password and click the Sign In button or click Sign In using your browser.

   When you click Sign In, all the dialog boxes close.

4. **Repeat Step 1 to reopen the preferences.**

   Your account with your avatar, name, and GitHub username appears under the GitHub.com row, confirming that you have successfully logged in.

5. **Click the Git tab.**

   The information has been autofilled for you and matches what you specified when you first set up GitHub Desktop.

6. **On the Appearance tab, choose Light or Dark.**

   Screenshots in this book are in Light mode.

7. **Set other preferences, such as the Editor and usage data, on the Advanced tab.**

   This book uses Visual Studio Code as the example editor, but you can select whichever editor you prefer.

   I recommend agreeing to send usage data, which is checked by default. By leaving it checked, you help the GitHub Desktop development team understand how all users use the application and therefore make it better.

   If you do not have a GitHub repository on your computer yet, you can stop the setup here. If you do have a repository, see Chapter 4.

FIGURE 2-7:
The Sign In dialog
box for the
GitHub Desktop
application.

REMEMBER

While a team of folks at GitHub predominately does the development of GitHub Desktop, a part of their role is to support community members who want to contribute to the project. Don't hesitate to reach out to the team on its repository at https://github.com/desktop/desktop.

Introducing Visual Studio Code

Visual Studio Code, also referred to as VS Code or even sometimes just Code, is a free, open source editor. Just like GitHub Desktop, VS Code is built on Electron, making it work on Mac or Windows PC. VS Code is *extensible*, meaning you can add your own features to it.

You can take a look at what the VS Code team at Microsoft is working on by visiting the repository: https://github.com/microsoft/vscode.

VS Code is a lightweight editor that shouldn't take long to install. To install it, go to https://code.visualstudio.com/ and click Download.

Just like with GitHub Desktop in the previous section, when VS Code finishes downloading, click to unzip the file. On Mac, the VS Code application appears in your Downloads folder. Drag the VS Code application into your applications folder. On Windows or Mac, double-click the application to open it. When you do, you should see what is shown in Figure 2-8.

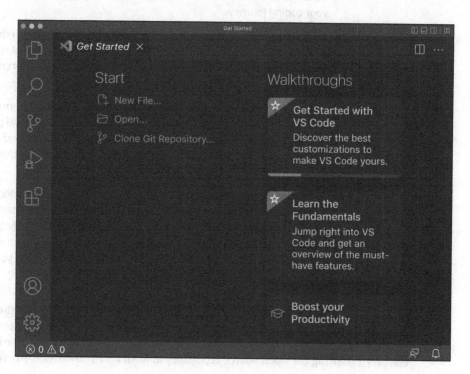

FIGURE 2-8:
The VS Code application default view.

WARNING

You may get an alert that you're trying to open an application that was downloaded from the Internet. Click Open if this alert appears.

Here are a few things that you should know about VS Code:

TIP

>> **Updates:** Each time you start VS Code, make sure you check the bottom- left corner to see whether any updates need to happen so that you keep your software as current as possible.

>> **Accounts:** At the bottom, left, above the settings cog icon, you find an Accounts icon where you can sign into your Microsoft or GitHub acocunts and sync your settings across computers all using a signed-in version of VS Code.

 You may notice that some menu items throughout VS Code have keyboard bindings. *Keyboard bindings* are combinations of keys you can press on your keyboard to make something happen in a specific application. You probably already know some of these. For example, when you're browsing the Internet, you can press ⌘-T on a Mac or Ctrl+T on Windows to open a new tab. Finding ways to become more efficient in your coding, such as by using keyboard bindings, can be an effective strategy for you as you become more expert in your coding journey.

>> **Preferences:** You can specify a lot of preferences for VS Code. I don't explain each one in this book, but I encourage you to browse them and really set up VS Code to make it exactly right for you. You can find the preferences by choosing Atom ⇨ Preferences.

>> **Command Palette:** Open the command palette in VS Code by pressing ⌘-Shift-P on a Mac and Ctrl+Shift+P on Windows. The command palette allows you to search for actions you can perform in VS Code, and suggests actions you likely want to take based on what you're currently doing in your editor.

>> **Extensions:** You can install more than 10,000 VS Code extensions to make VS Code most effective for you. You can find them under extensions or at https://marketplace.visualstudio.com/VSCode. Going through each of these extensions is beyond the scope of this book, but I encourage you to explore some and search them if you're ever feeling limited by your VS Code experience.

REMEMBER

If you're looking for resources outside of this book, check out the VS Code documentation at https://code.visualstudio.com/docs. If the docs don't help you resolve your issue, you can reach out to the developers who work on VS Code by visiting the open source repository at https://github.com/microsoft/vscode.

2

Starting Your
First Solo Project

Chapter 3

Introducing GitHub Repositories

Almost everything on GitHub.com revolves around a repository.

In this chapter, you find out how you can set up a repository, interact with it, and create project boards and issues. The repository that you set up in this chapter is a special kind of repository. The functionality of the repository is the same, except that this repository is named the same as your GitHub username, which makes it automatically appear on your GitHub public profile, such as mine at https://github.com/drguthals. Adding information to your public profile acts as a cover page for your software profile, gives you control over how people understand who you are and what you do, and helps enable trust when you are contributing to open source projects.

Setting Up a Repository

A GitHub *repository* (or *repo*) is a folder with all the files needed for your project, including the files that track all the versions of your project so that you can revert any mistakes you make. A repository on GitHub also tracks who can collaborate and how.

To get a better understanding of what a repository is and how it's structured, you need to create your first GitHub repo:

1. **Go to the home page of GitHub.com; if you're already on GitHub, click the Octocat to get to the home page.**

   A list of your repositories appears on the bottom-left side of the screen.

2. **Click the green New Repository button.**

   The Create a New Repository page, shown in Figure 3-1, opens.

3. **Type the name of your repository in the Repository Name text box.**

   Make sure you name this repository the same as your GitHub username, including casing and any other characters. In my case, I named my repository dra–sarah.

4. **Select the Public radio button.**

   This repository needs to be public so that it can be viewed on your public GitHub profile page.

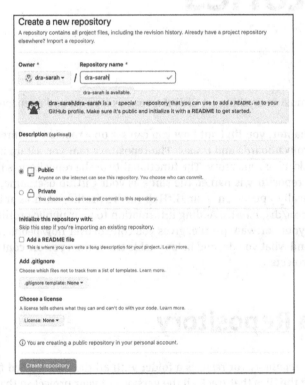

FIGURE 3-1: The page to create a new repository.

5. **Under the Initialize This Repository With section, click the Add a README file check box.**

   You do not need to add a `.gitignore` file. This is used to make sure you do not add local files (such as cache files) to your repository, but because this repository is only a README file, you do not need a `.gitignore`.

6. **Choose a license from the Add a License drop-down list.**

   If you're interested in finding out more information about licenses, see the nearby "Software licenses" sidebar.

TIP

7. **Click Create Repository.**

   The home page of your repository appears. It should look similar to the one I created, which is shown in Figure 3-2. Notice that a Markdown file — `README.md` — is already in the repository. Markdown is a lightweight markup language used to style the words that you write with a plain text syntax. You can make words bold, turn them into headers, and even create a table for data.

   Now you can head back to your profile page, for example the one I just created would be at `https://github.com/dra-sarah`, and see your README displayed at the top.

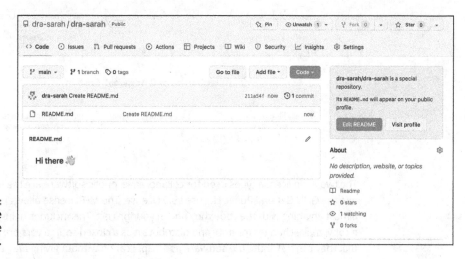

FIGURE 3-2:
The home page
of my profile
repository.

In Chapters 4 and 5, you can create a website for yourself. This website can link back to your repository.

SOFTWARE LICENSES

Software licenses are a really important part to collaborative coding. Whether you're putting your code up on GitHub.com to share with the world, or contributing to someone else's code, you should know what is allowed and what isn't. When you first create a new repo, you're given the option to attach a license to it (refer to Figure 3-1). If you click on the question mark, you will be taken to https://chooselicense.com/, which explains the top three most common software license types, and shown in the following figure.

The two main license types used for collaborative, public software are the MIT License and the GNU General Public License v3.0 License. The MIT License allows people to do almost anything with the code they find in your project. This includes being able to take a copy, make changes (or not), and distribute it as a closed source version. That means that they can distribute the software as an application without giving *their* users access to the code. The GNU General Public License v3.0 License also gives folks access to copy, modify, and contribute to your project, but if they want to distribute their version, it must be public and open. The repositories in this book use the MIT License, but you can choose what you want to do with your projects.

Exploring Your Repository

A repository has a lot going on, even when it's as simple as the one one that I created in the preceding section (refer to Figure 3-2). The following sections walk you through an overview of everything on the repository.

Top information

At the top of the repository is the username of the author and title of the repository. When you fork a repository, you see the original author underneath for a quick link. To *fork* a repository is to make a copy of it, where the changes you make to your copy can be suggested to the original author. See Chapter 6 for a deep dive into forking a repository.

To the right of your username are three buttons:

>> **Watch:** You can choose what kind of notifications you want to receive based on the type of activity happening on this repo.

>> **Fork:** If you're not the author of the repository, then you have the option to fork it. Chapter 6 goes into more detail on forks.

>> **Star:** Starring can help you quickly navigate to certain repositories, as well as give GitHub insight into things you're interested in so that recommendations are more accurate for you. To access your starred repositories, just click your avatar on the top right of GitHub.com and choose Your Stars.

WARNING

When choosing to watch a repository, I highly recommend choosing either Ignore or Participating and @mentions for the majority of repositories so that you only get notifications when you're specifically mentioned or actively participating in a discussion on an issue or a pull request. Otherwise, your inbox fills with emails about every single action taken on the repository, which can get out of hand very quickly. If you notice this happening, go to `https://github.com/watching` and unwatch all or some of the repositories you're watching with a quick click.

Tabs

Nine tabs appear across the top of your repo. Each tab provides different features for the repo:

>> **Code:** The Code tab is where you can find all your code and browse folders and files. You can click a file to view its contents or click the pencil icon to

modify the file, right in your Internet browser. (See the upcoming "Code tab" section for more details.)

>> **Issues:** Issues are a really neat feature for repos. Issues can help you track things you want to still make, problems you're having, or suggestions for other people. You discover how to create issues in the upcoming section "Using Issues and Project Boards."

>> **Pull requests:** Pull requests, also referred to as PRs, are similar to issues in that they have a title and a description, but they also have code changes that you're requesting to be pulled into the main branch. The safest way to contribute code is to create a new branch, make your code changes on that branch, and then request for that branch to be merged with the main branch. A PR gives you an interface for merging the two branches, showing you the *diff* (list of changes or differences) between the files you modified and the ones that are on the main branch and giving you a place to have a conversation with collaborators on whether the code should be merged or changes should be made first. For more information on branching, see Chapter 1.

>> **Actions:** GitHub Actions is an integrated way to introduce automation with your repository. Actions are often used to build, test, and deploy your code directly from your repository. You can create a number or automations; learn more at https://github.com/features/actions.

>> **Projects:** You may already be familiar with project boards like Trello or Kanboard. GitHub has project boards linked directly with your repo. The best part is that the cards on a GitHub project board can be directly related to issues or PRs and can automatically move when something happens. For more on project boards, see the section "Using Issues and Project Boards," later in this chapter.

>> **Wiki:** Wikis are a great place to store documentation, project status, and roadmaps for your project. It's a great go-to place for collaborators to see what is going on and where they can jump in to help!

>> **Security:** The Security tab gives you an overview of how others should report security vulnerabilities on your repository, a place for you to list security vulnerabilities you're aware of, dependabot alerts, and automated code scanning alerts.

>> **Insights:** The Insights tab, shown in Figure 3-3, gives you an overview of all collaborators and actions happening on the repo. It's really neat to see this tab on popular open source projects. For example, TensorFlow has had 249 contributors in the last month!

>> **Settings:** The Settings tab is visible only if you have the right permissions on the repository. In this tab, you can decide who has access to what and how collaborators should collaborate. You can also integrate apps that tell you how much of your code is covered with tests.

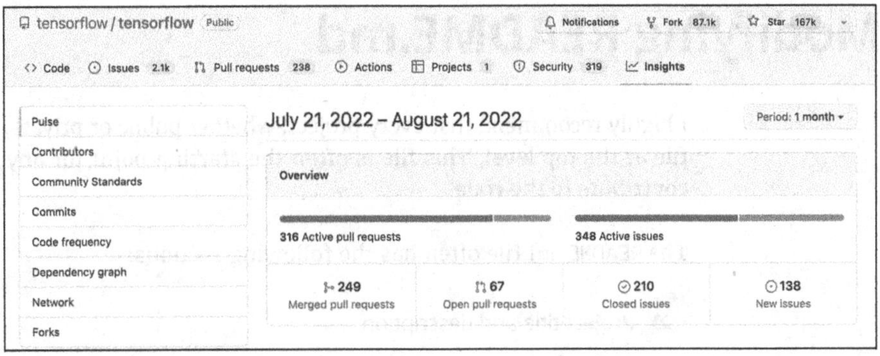

FIGURE 3-3:
The Insights tab.

Code tab

The Code tab has a lot of additional important metadata about your repo that will come in useful in future development:

>> **Description and topics:** At the top-right of the Code tab is a description and a place where you can put in topics to make your repository more discoverable. Adding topics is particularly important if you want to attract other coders to help you build your software.

>> **Metadata:** On the right under the description of the Code tab you find information and links to licenses, number of stars, people watching, and forks. At the top of the Code tab you find additional information to the number of branches, and number of tags on the repository.

>> **Action buttons:** On the top of the Code tab is a drop-down menu where you can change which branch you're looking at or browse the files for a particular branch. The New Pull Request button appears if there is a branch that is out of sync with the main branch and allows you to quickly create a pull request. The best way to create a pull request is to switch to another branch, make some changes, and then click New Pull Request. You might find other information just above your list of files, for example a suggestion to protect your main branch if you haven't already. On the right side are three buttons related to files: Go to File, Add File, and Code. Clicking Go to File takes you to a new page where you can search for a specific file within the repository. Clicking Add File gives you two options: Create New Fle, Upload Files. Finally, you can click the green Code drop-down list to clone or download the code to your local machine (see Chapter 4). Here you also have a tabbed option to open the code in a Codespaces instance.

>> **Code:** At the bottom of the Code tab is a list of all the code in this repo. If a README.md file appears in this list, then the file shows up below the list. For any file, you can click the filename to go to a page where you can see the file and edit it if you want.

Modifying README.md

I highly recommend that every project, whether public or private, have a README.md file at the top level. This file is often the starting point for anyone who wants to contribute to the code.

The README.md file often has the following sections:

- » Project title and description
- » Prerequisites for getting the project running on your local machine
- » Instructions on installing the project (and any dependencies)
- » Instructions on running tests to make sure that you haven't broken anything
- » Instructions on deploying the project
- » An overview of dependencies
- » A link to the guide on how to contribute to the project, including a code of conduct
- » The main authors or maintainers of the project
- » A link to the license
- » Any additional acknowledgements

TIP

PurpleBooth on GitHub has created a great template for a README.md file at https://gist.github.com/PurpleBooth/109311bb0361f32d87a2.

REMEMBER

GitHub promotes a culture of sharing and open software development. In the sharing, it's important that each person acknowledge where they drew inspiration and what pieces went into helping them create what they have created. Software development rarely happens alone and, at this point, is always built on someone else's work. Though you don't have to specifically acknowledge the work that Grace Hopper, a well-known computer scientist who created the first compiler and English programming language, did to promote high-level programming languages so that you're not all writing in assembly anymore, you should always recognize that it's a large, timeless community working toward building, creating, and pushing the boundaries of what you think is possible today.

For simpler projects, a README.md file can also be the front page to your project.

For this special type of repository, the README.md file is what's displayed on your GitHub profile page, so it's a little different than a typical repository. Good news is, modifying the README.md file is the same for this and all repositories.

Before you jump into creating your `README.md` for your public profile, check out some examples of what other people do:

>> My profile has links to other content channels and an overview of my career: `https://github.com/drguthals`.

>> Phil Haack has a fun gif, links to social media, and his GitHub stats: `https://github.com/haacked`.

>> Brian Douglas, a developer advocate at GitHub, has re-created the MySpace "Top 8," including Tom: `https://github.com/bdougie`.

>> April Speight, a developer advocate at Microsoft, has amazing cover art, links to her publications, and ways to connect with her: `https://github.com/aprilspeight`.

>> Jordan Harband, an avid open source maintainer, has a list of projects he maintains, standards he contributes to, and GitHub stats: `https://github.com/ljharb`.

Even if you don't know exactly how you want to have your profile `README.md` yet, you can follow these steps to modify it with a bit about how you got started:

1. **Start by creating a new branch to work off of by clicking the branch drop-down menu shown in Figure 3-4 and typing the name of a new branch.**

    I named my new branch `initial-profile-readme`.

FIGURE 3-4:
The branch drop-down menu on a GitHub repository.

2. **Click Create branch:*yournewbranch*-readme from 'main'.**

   The branch for your repository is now listed in the drop-down menu and selected. You can move between this branch and the main branch through this menu.

   This step allows you to see your project as it looks in that branch. The changes you make are added to that branch as you make them. This is particularly handy when you're intending on making changes to more than one file.

3. **At the top right of your README file, click the little pencil so that you can change what the README says.**

   If it exists, a README.md file always appears below the list of files.

4. **Using Markdown, write a little about yourself, including your career passions and some hobbies you enjoy.**

   Remember, you can be creative here. In fact, GitHub suggests checking out the Emoji Cheat Sheet: https://www.webfx.com/tools/emoji-cheat-sheet/ because you can put emojies in by using their code :tada:.

TIP

   When you begin to edit the README there is a Markdown comment that isn't displayed in the published version of the README, but gives you an idea of where to start. Figure 3-5 shows this Markdown comment in the GitHub editor. Notice that if you click the Preview Changes tab above the text, the text within the comment brackets (<!-- -->) is not displayed.

5. **When you're satisfied with your README, scroll to the bottom of the file editor, add a title to the commit (grouping of modified, created, and deleted files), and commit the changes to the same branch you just created.**

   You see your README.md file in its final state.

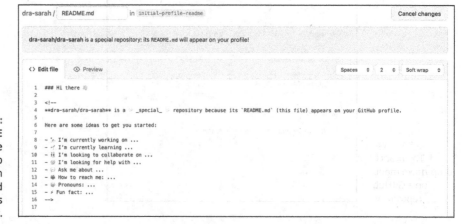

FIGURE 3-5: The README Markdown file in the GitHub editor with the suggested changes comment.

```
dra-sarah /   README.md        in initial-profile-readme                                    Cancel changes

dra-sarah/dra-sarah is a special repository: its README.md will appear on your profile!

<> Edit file    Preview                                              Spaces  ⬦  2  ⬦  Soft wrap  ⬦

  1   ### Hi there 👋
  2
  3   <!--
  4   **dra-sarah/dra-sarah** is a _special_     repository because its `README.md` (this file) appears on your GitHub profile.
  5
  6   Here are some ideas to get you started:
  7
  8   - 🔭 I'm currently working on ...
  9   - 🌱 I'm currently learning ...
 10   - 👯 I'm looking to collaborate on ...
 11   - 🤔 I'm looking for help with ...
 12   - 💬 Ask me about ...
 13   - 📫 How to reach me: ...
 14   - 😄 Pronouns: ...
 15   - ⚡ Fun fact: ...
 16   -->
 17
```

6. **Click the repository link at the top where it says** YOUR_USERNAME/ README.md **to return to your code tab.**

   Notice the suggestion to compare the changes on this branch with the main branch, and to create a pull request, shown in Figure 3-6.

   This time is perfect to open a *pull request*, a request to save the changes you have made to the branch you're targeting. You can always make additional changes before you merge the pull request, but creating the pull request now helps you track the changes easier.

FIGURE 3-6: GitHub's suggestion to open a new pull request when a new branch is created.

FIGURE 3-6: GitHub's suggestion to open a new pull request when a new branch is created.

7. **Click the Compare and Pull Request button.**

   In the screen shown in Figure 3-7, you can create a pull request. While this initial pull request is fairly simple, it's always a good idea to include a description. I'll add the following as mine:

   ```
   Initial README for my [profile](https://github.com/dra-sarah) that links
       to my GitHub For Dummies book, my actual GitHub profile, and includes
       some facts about me.
   ```

8. **Under the pull request description box, click Create Pull Request.**

   Once the pull request is created, if you make edits on the same branch (in this example, the initial-profile-readme branch), those edits are automatically included in this pull request.

You have now made changes to your project. The only problem is these changes are still on their own branch, and not on the main branch. To find out how to merge your changes into the main branch, see the next section.

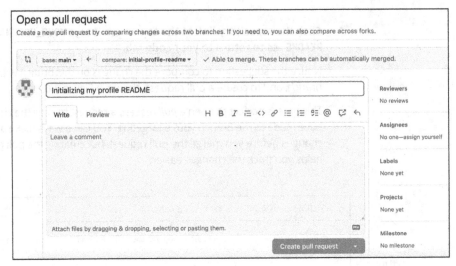

FIGURE 3-7:
A pull request
edit view.

Merging a Pull Request

After you have all your changes in a pull request (see the preceding section), you can merge those changes into the `main` branch by following these steps:

1. **On the home page, click the Pull Requests tab, and then select the pull request you want to merge, in this case #1, to get to your pull request details.**

Figure 3-8 shows the Pull Request List page while Figure 3-9 shows the open pull request.

FIGURE 3-8:
My pull
request page.

2. **Click the Files Changed tab to see all the changes made to this repo, as shown in Figure 3-10.**

   Files and lines that appear in red will be deleted, while the files and lines in green will be added.

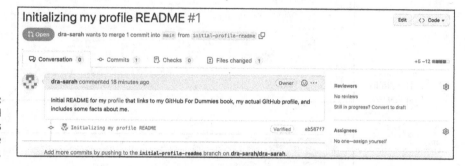

FIGURE 3-9:
The Pull Request Details page for a single pull request.

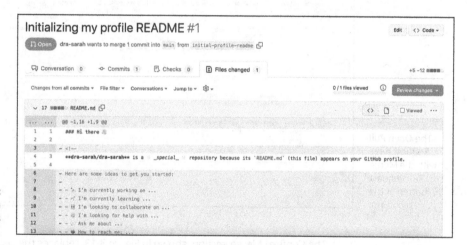

FIGURE 3-10:
The diff shown on the Files Changed tab.

3. **(Optional) To change the way you see the diff, click the Diff View dropdown menu and then click Split, and then click Apply and Reload.**

   If you split the view, your screen changes (see Figure 3-11).

4. **In the Conversation tab, scroll to the bottom of the pull request and click the big green Merge Pull Request button, as shown in Figure 3-12.**

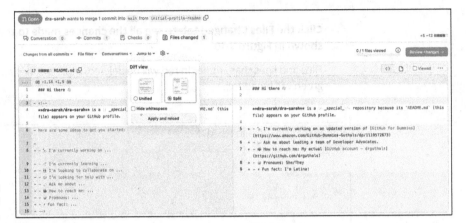

The Confirm Merge section, shown in Figure 3-13, replaces the section with the Merge button.

5. **Click Confirm Merge.**

You see a message that your pull request merge was successful, with an option to delete the branch (see Figure 3-14).

6. **Click Delete Branch.**

Your pull request is merged, and the branch is deleted. Don't worry, if you need that branch back for some reason, you can restore it. It's nice to keep things tidy within the repository.

7. **Click the Code tab to go to your code.**

You see the main branch with your picture and the changed README.md file.

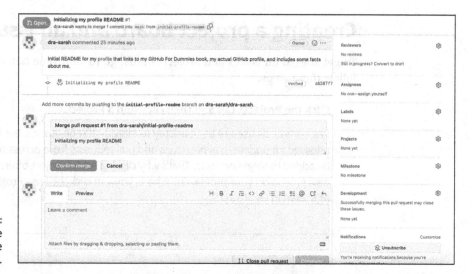

FIGURE 3-13:
Confirm the
merge of the
pull request.

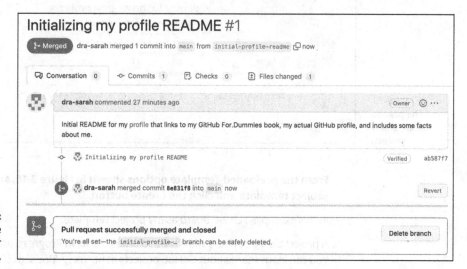

FIGURE 3-14:
The merge
confirmation for
a pull request.

Using Issues and Project Boards

Issues on a GitHub repo are a great way to track the things you need to fix, add, or change. When you combine issues with project boards, you get insights into your project that would otherwise be hard to track. In this section, you create issues and project boards and change your README.md.

Creating a project board and an issue

To get started on issues and project boards, go to your repo home page and then follow these steps:

1. **Click the Projects tab and then click Add a Project.**

Project boards used to be restricted to individual repositories, but GitHub has released an update to allow issues and pull requests from across repositories be added to single projects. That's why clicking Add a Project offers to create a new project on your profile. From here, you can click the New Project button, as shown in Figure 3-15.

FIGURE 3-15:
The page to create a new project.

2. **From the preloaded Template options shown in Figure 3-16, select a project template and click the Create button.**

In this example, I chose Board as my project template.

A project board appears with three columns: Todo, In Progress, and Done. You can learn more about GitHub projects at `https://docs.github.com/issues/planning-and-tracking-with-projects/learning-about-projects/about-projects`.

3. **Click the + button at the bottom of the Todo column.**

Here you can either type an issue number to add an existing issue from your repository to the Todo column in this project, or you can create a draft issue by typing a short sentence to describe a new task you want to keep track of in this project. A draft issue can be converted to an actual issue tracked on your repository after you create the draft issue card on the project. Create a new draft as shown in Figure 3-17.

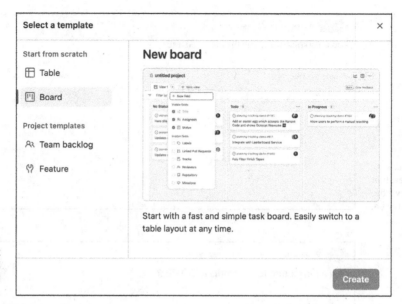

FIGURE 3-16:
The project
template options
with the board
option selected.

FIGURE 3-17:
A drafted issue
being added to
the Todo board.

4. **Click into the draft card and choose Convert to Issue, as shown in Figure 3-18.**

Once you choose which repository you want to create the issue in, the draft view refreshes and a preview of the issue replaces it, as shown in Figure 3-19. You can click the Open button to see the issue in the repository's Issues list.

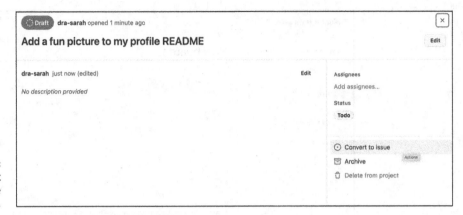

FIGURE 3-18:
The view that
shows the
drafted issue.

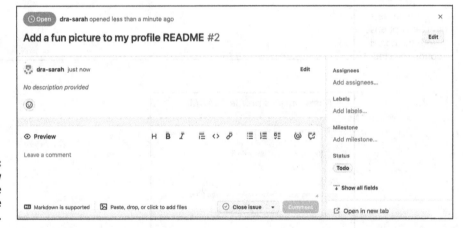

FIGURE 3-19:
The view
that shows the
issue in the
project board.

Closing an issue

The best way to close an issue is to create a pull request with changes to address what was written in the issue. Understanding the relationship between issues and pull requests can help you on your own projects and open source projects.

To close out the issue, follow these steps:

1. **From the Code tab on your repo, click the pencil icon for the README.md file to edit the file.**

2. **Add an image at the top of the README.md file with your favorite animal.**

   To do this, drag an image into the editor. The image is uploaded to GitHub and the following line of code is added to the README (though your exact user content is different):

```
![axolotl](https://user-images.githubusercontent.com/110885554/186362310-
    d5c6e677-5b67-49b2-bf77-e0f41e5b1949.png)
```

3. **Scroll to the bottom of the page and add a title to the commit.**

4. **Choose Create a New Branch for this commit and start a pull request.**

5. **Click Commit Changes.**

6. **Add a description to the pull request.**

   Specifically, make sure that you write `closes #2` on its own line. When you type **#**, GitHub suggests any issue or pull request that you have in this repo to autofill.

7. **Click Create Pull Request.**

   At the bottom right of your pull request is a Development section that references the issue you reference in the pull request description.

8. **Click through to open this issue, and then click into the project board that the issue is a part of to view the issue in the Todo column.**

9. **Click the Back button on your browser twice to get back to the pull request.**

10. **If you're happy with the changes, click the Conversation tab, click Merge Pull Request button, click Confirm Merge button, and then click the Delete Branch button.**

    You can revisit the section "Merging a Pull Request," earlier in this chapter, for more details on how to complete this step if you get stuck.

11. **Click the Issues tab to find the issue now marked as closed.**

12. **Click the Closed tab on the Issue page and click the issue that was closed. Navigate to the project board where this issue was in the Todo column and notice it has automatically moved to the Done column.**

Chapter **4**

Setting Up a GitHub Website Repo

One technology that truly pushed the boundaries of society and software development was the Internet. As the Internet became more embedded in everyday life, it brought new meaning, career opportunities, and ways to connect for millions of people. A more recent phenomenon enabled by the Internet is a set of sites categorized as social media. Social media profiles provide a way for people to express who they are, what their interests are, and how to connect with them. But social media creates new challenges.

Personally, I have accounts on Twitter, Instagram, TikTok, Polywork, and LinkedIn. I don't feel comfortable sharing personal accomplishments (like having a baby) on my LinkedIn, but also rarely share minor professional accomplishments (like starting a new project) on my Instagram. For me, having a website gives me a place to point people to all of me, not just the version of me that fits the community of the particular platform I'm using. That way, if folks want to learn more about the new project I'm working on, they will head to my website where they may also stumble upon the fact that I have a podcast with Chloe Condon about movies and shows from the '90s. This interaction improves my bonds and connections with people from all aspects of my life.

Others may use a custom website to still focus on a particular part of who they are, but they like having the control over what they share and how they do it. This chapter, along with Chapter 5, guides you through turning any project repo that you own into a website and creating your own website on GitHub.com. In a matter of minutes, you can have a website up and running without having to pay for additional services.

A *repo*, or *repository*, is a coding project contained in a single folder where modifications to files is tracked. While you can have a local repository that only exists on your local computer, this book is typically referring to the hosted repository on GitHub.com, unless otherwise stated.

Introducing GitHub Pages

GitHub Pages is a fast and easy way to make a website that is hosted on GitHub. com. The code in your repo will be the code running the website. Even better is that it's much easier to style your websites with Jekyll, a free, open source site generator that takes Markdown files and creates websites with support for themes.

TIP

You can discover more about Jekyll at https://jekyllrb.com or even check out what it's up to on its GitHub repo (https://github.com/jekyll/jekyll).

With GitHub Pages, you can create a website using Markdown or HTML/ JavaScript/CSS.

REMEMBER

If you need help remembering what you can do with Markdown, visit the Markdown GitHub Docs at https://guides.github.com/features/mastering-markdown.

Turning a Project Repo into a Website

GitHub Pages is a great tool that is integrated into GitHub.com. GitHub Pages looks for a README.md file on your main branch and use it as the landing page for your website, meaning you don't have to do much to get it up and running! Just follow these steps:

1. **Open a repository on GitHub.com.**

   For this example, I created a simple HelloWorld repo that has a basic README and a *very* simple solution for a sum function written in Python. You can see this repository at https://github.com/dra-sarah/HelloWorld.

2. **On the home page for the repo, click the Settings tab on the top right to open the Settings page.**

3. **Scroll down on the left menu and click the Pages option, shown in Figure 4-1.**

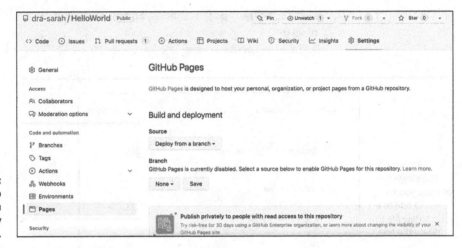

FIGURE 4-1:
The GitHub
Pages section
under repository
settings.

4. **From the Source drop-down menu, change the source for GitHub Pages to Deploy from a Branch. From the Branch drop-down menu, change the branch to deploy from None to** `main` **and then click Save.**

5. **Choose a theme for your website.**

   Under the specified deployment branch you should see a link to choose a theme, shown in Figure 4-1. The Jekyll theme docs page opens, where you can browse supported themes at `https://pages.github.com/themes/`.

6. **After you choose a theme for your website, go back to the Code tab on your repository to add the config file.**

   Click the Add File drop-down menu and select Create New File, as shown in Figure 4-2. Name the file `_config.yml`.

FIGURE 4-2:
The GitHub add a
new page option
in the repository.

7. **Add the theme you chose to the config file, shown in Figure 4-3. Then commit the changes to the** main **branch by clicking the Commit Changes button at the bottom of the editor.**

The config file should really only have one line of code; in this example, I have chosen the theme time-machine:

```
remote_theme: pages-themes/modernist@v0.2.0
plugins:
- jekyll-remote-theme
```

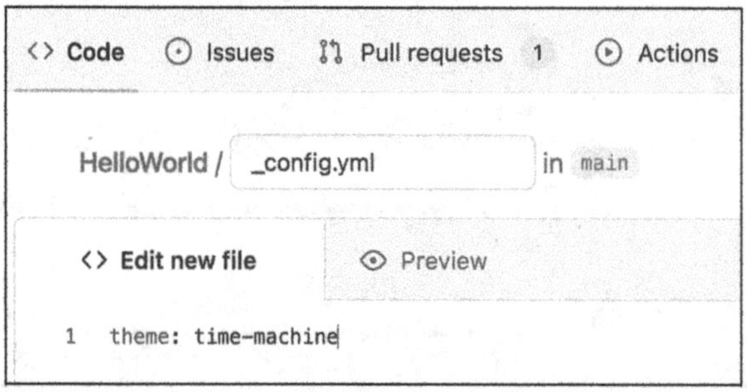

FIGURE 4-3:
The single line
of code you have
to add to a
single file to
update the title.

8. **On the Code tab, click the the Environments option on the left menu, as shown in Figure 4-4.**

9. **Click the active GitHub Pages environment, and then click the View Deployment button on the right-hand side.**

Now you see a web page that says Hello World, just like is written in the README file for your repository.

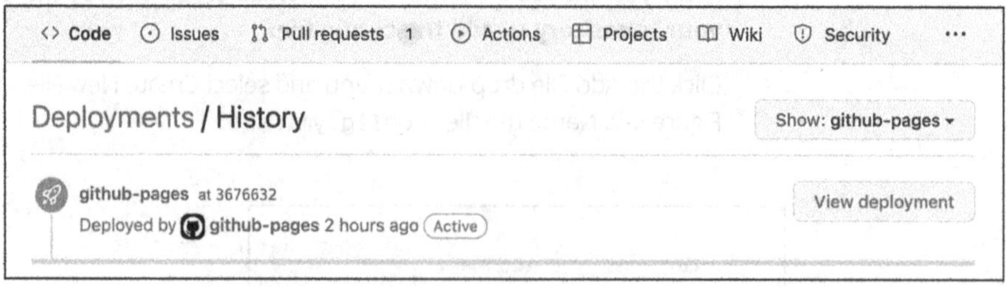

FIGURE 4-4:
The GitHub Pages
website theme
chooser.

Setting Up a Personal Website Repo

To create a new repo that houses your own personal website, you need to set up your repo:

1. **Create a new repository and name it** *username*.github.io, **where** *username* **is replaced with your actual GitHub username.**

   For example, the name of my repository is dra-sarah.github.io. If you're unsure how to create a new repository, see Chapter 3.

2. **Make the repository public, initialize it with a README, choose a license, if you want, and then click Create Repository.**

   If you're unsure whether you want a license, see Chapter 3, which describes the benefits of choosing a license.

   The page refreshes to the home page of your new repository.

3. **Create a new project that is the board style.**

   If you don't know how to create a project, see Chapter 3 for guidance.

4. **Once you have created the project, head back to the project page of your repository, click the Add Project drop-down menu, and choose the project you just created, as shown in Figure 4-5.**

   Now this project has a spot in the Quick Access section, shown in Figure 4-6.

5. **At the top of your repo again, click the Settings tab and scroll down until you see the GitHub Pages section.**

FIGURE 4-5:
Adding a project to a GitHub repository for quick access via the Projects tab.

FIGURE 4-6:
A GitHub pages
project linked to a
specific
repository.

Notice that it says your site is ready to be published at a certain URL (mine is https://dra-sarah.github.io/). If you go to this URL, you see a simple barebones web page with the contents of your README.md file. If you get a 404, just wait a moment and refresh the page. It can take a few seconds for GitHub Pages to build your site.

6. **In the GitHub Pages section, click Choose a Theme and then click Select Theme.**

   If you don't know how to do this, review the previous section in this chapter, "Turning a Project Repo into a Website."

7. **Edit and commit your README.md file.**

   Back on the Code tab of your repository, click the pencil icon in the top-right of your README.md file and add some information about you. Commit those changes to a new branch and open a pull request. The page refreshes with a new pull request ready to be created. You can merge your pull request. Then you can navigate back to your code, and you see the _config.yml file and the changes to your page.

8. **Create a pull request with a more descriptive title.**

   For instructions on creating a pull request, see Chapter 3.

   Change the title of the pull request to describe everything you want to do for this iteration of your website and add a description. For example, I added the following list:

```
To create a basic website:
- [ ] Add a short personal bio
- [ ] Add links to my social media channels
```

9. **Link this pull request to the project and update the status of this item.**

To ensure your project board automatically tracks the progress of your website, choose the project created in Step 3 on the right side of the pull request from the Projects section. Then update the status of this item to "In Progress," shown in Figure 4-7.

FIGURE 4-7:
A GitHub pull request linked to a specific project with a status of "In Progress."

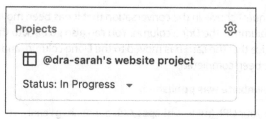

10. **Update the pull request when appropriate.**

TIP

The description of a pull request is not static. When you first create a pull request, you may have a list of things you want to do before merging the code into the `main` branch. You may also end up making changes throughout. Make sure to always revisit the description and make sure it is accurate. For example, if you're following these steps exactly, you have already added some information about yourself and maybe even links to your other social media channels, so you can likely check off those two boxes.

11. **Verify the project board automation.**

Go back to the project board by clicking the Projects tab at the top of your repo and notice that a new card is in the In progress column. If you click into the card, you're redirected to the pull request.

12. **Switch to the pull request branch.**

Go back to the Code tab of your repo and switch to the branch that is associated with your pull request. Mine is `readme-init`.

TIP

If you don't know how to switch to a different branch on GitHub.com, see Chapter 3.

13. **Create an `index.md` file.**

On the top right of your code file list, click Create New File. Name the file `index.md` and add a header that says "Hello World!":

```
# Hello World!
```

Commit that file to the same branch you've been working on, with a title and commit message.

14. **Update and merge the pull request.**

Go back to the pull request and add a new item to your checklist "create an index.md landing page," and then check it off. Because everything that you wanted to do for this pull request has been completed, the pull request can be merged. Click Merge Pull Request, Confirm Merge, and then Delete Branch.

15. **Verify the project board automation.**

The pull request shows in the conversation that it has been moved from the In Progress column to the Done column. You can also go back to the project board. Notice that the card has moved to the Done column, and both checklist items have been completed.

16. **Verify the website was published.**

Go back to your URL (mine is `https://dra-sarah.github.io/`). You have a working website with the theme you chose.

TIP

If you don't see the changes you've made to your website, like the Hello World message or the theme that you chose, try refreshing your web page. You now have a website that you can continue building and customizing as you do more and have more to share with the world.

Creating Issues for Your Website

After you have a GitHub.com website repository (see preceding section), you can think through the sections you want to have on your website. Creating issues for everything you want to add or change about your website can help you plan and remember all the little things you want to change.

Say that someone gives you a great suggestion. You don't want to pull out your computer and make the change right then and there, but you can quickly jump on GitHub.com and create an issue to remind you to add it later. Creating an issue can also be useful if you are working on your website, and you've found something that is bothering you that you want to change. Instead of derailing what you're already working on with a new task, you can just make a quick issue and get to it later.

To get started with this planning phase, go to the Issues tab on your repo and click New Issue. Create an issue for all the things you want to add to your website.

In this book, I created two issues to use in my example:

» **Change the title and tagline.** The title and tagline of your website is currently something auto-generated. You probably want to change it to your name and some tagline that represents you. Create an issue, assigning yourself to it and linking it to your project board:

```
Issue Title: Change the title and tagline
Issue Description: Make the title and tagline something
    unique to me.
```

» **Add sections to the website.** Without even having to leave an issue you created, you can click New Issue and create another issue to add in sections to your website, assign yourself to the issue, and link it to your project board. In my example, I've chosen to add three sections:

```
Issue Title: Add a couple of section]s to the website
Issue Description: Add three sections to the website:
- [ ] About Me
- [ ] Contact Information
- [ ] Current Focus
```

Two issues now appear under the Issues tab of my website repo, as shown in Figure 4-8. They also appear on my project board in the Todo column (see Figure 4-9).

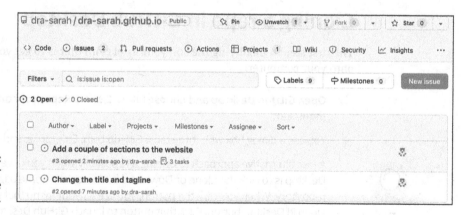

FIGURE 4-8:
The issue list for
the website
repository.

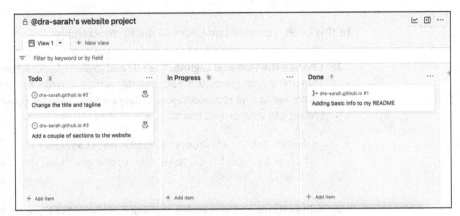

FIGURE 4-9:
The project board
for the website
repository.

Setting Up Your Local Environment

This section assumes you already set up GitHub Desktop and Visual Studio Code. If you haven't, Chapter 2 can help guide you through this process.

In this section, you get your website working so that you can modify files on your local computer instead of on GitHub.com.

TIP

Modifying files on your local computer can be useful if you need to work on a project when you won't have Internet access or if you need to browse a lot of files while editing files.

Cloning a repo in GitHub Desktop

The first step in modifying files on your local computer is to get your website repo onto your computer:

1. **Open GitHub Desktop and choose File ⇨ Clone a Repository on the menu bar.**

    You see a dialog box with three tabs: GitHub.com, Enterprise, and URL.

    TIP

    A nice alternative approach to cloning a repository when you have GitHub Desktop is to click the Clone or Download button on the home page of every repository. When you click the button, you see a flyout menu that includes an Open in Desktop button. Click that button to launch GitHub Desktop (if it's not already running) and clone the repository to your local machine.

2. **On the GitHub.com tab of the Clone dialog box, your repositories autofill for you from GitHub.com.**

WARNING

If your GitHub.com repositories don't autofill in the Clone dialog box, it probably means you're no longer signed in to GitHub.com in GitHub Desktop. You can log in by choosing GitHub Desktop ⇨ Preferences from the top menu bar. Click the Accounts tab and sign in. If it appears you're logged in but your repos are still not showing up, try signing out and signing back in.

3. **Choose your personal website repository and choose where you want it to be stored on your local machine.**

   I chose the default path as the place to store the repository on my local machine.

4. **Click Clone.**

   GitHub Desktop refreshes with your repo information included.

Touring GitHub Desktop

GitHub Desktop offers a variety of features to help you with your development and interactions with GitHub.com. You can check out the GitHub Desktop User Guides at `https://help.github.com/desktop` if you need additional support beyond this book. Figure 4-10 highlights the top six features:

>> **Repository list:** As you clone more repositories to your local computer, clicking the Current Repository drop-down menu reveals all the repositories that you have on your local computer, enables you to quickly switch between them, and gives you a button to quickly add a new one.

>> **Branch list:** The branch list gives you a quick overview of all the branches that you have checked out on your local computer, as well as a button to quickly create a new branch.

>> **Pull request list:** One the same drop-down list as the branch list, you see a second tab that lists all the pull requests that are open on this repo.

>> **Sync Project button:** As you start to make changes and/or changes are made on the repo outside of what you're doing on your local machine, you need to sync. Because Desktop hasn't detected any changes made on GitHub.com or your local computer, the option presents itself as a fetch to start. If you start to make changes on your local machine, you can choose to push your local changes to GitHub.com. If you start to make changes on GitHub.com, you can choose to `pull` those changes to your local machine. If you create a repository on your local machine and it isn't on GitHub.com yet, you can choose to publish your project to GitHub.com.

WARNING

If you do not push your changes to GitHub.com, they won't be available for other people and if your computer were to crash, you would lose all your work. I highly recommend that you push your code often.

>> **Changes list:** As you start to make changes to your code, the files that you've added, deleted, or modified show up in this changes list. You can click each file to see the diff to the right. When you're ready to commit to those changes, you can add a Summary and Description and click the Commit to Main button. At that point, you can push your changes to the branch that you're on clicking the Sync Project button.

WARNING

You should always double check which branch you're on before you commit and push your changes. You can undo commits and pushes, but avoiding it is best because the process can get hairy really quickly.

>> **History list:** Next to the changes you find the history of this repo. The history includes activity from your local machine that has been synced with GitHub. com and activity from GitHub.com that you may have never done on your local machine. When you click one of the events, you see a list of activity that happened.

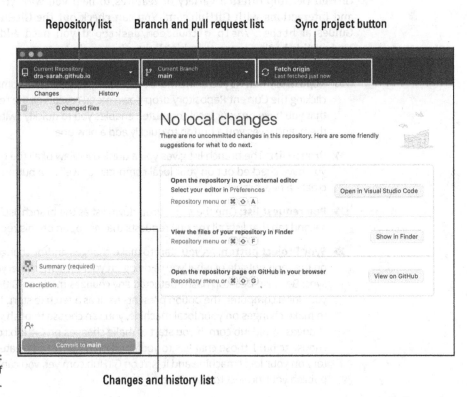

FIGURE 4-10:
An overview of
GitHub Desktop.

Opening your repo in Visual Studio Code

To edit your files on your local computer, you can use a number of applications, including Visual Studio Code or even TextEdit. In this book, I use Visual Studio Code (VS Code).

To open your repo in VS Code:

1. **Open VS Code.**

   You see a blank window.

2. **Choose File ⇨ Open from the top menu bar.**

   A file finder dialog box appears.

3. **From the file chooser, open the folder for your repo and click Open.**

   Your project is now open in VS Code.

Touring VS Code

VS Code is primarily a code editor, but it also has features that make coding on GitHub repos much easier. If this section doesn't offer enough detail for everything you can do with VS Code, make sure that you check out the docs (https://code.visualstudio.com/docs). In particular, the VS Code repository is always updated to reflect the newest features and what the team is working on. The Issue page of the repository typically has a monthly iteration plan for even more detail (https://github.com/microsoft/vscode/issues). Figure 4-11 shows the top six features:

>> **File list:** On the left side of VS Code is a list of all the files that you have in your repo. If you click a file, it opens in the center code editing area.

>> **Code editor:** To the right of the file list is the code editor where you can write or modify code.

>> **Branch list:** At the bottom left corner of VS Code is a branch chooser. Right now, you are on the main branch of your repo. If you click the branch, a menu opens, allowing you to choose between branches or create a new one.

>> **Sync Project button:** VS Code supports syncing your project with what is on GitHub.com.

>> **Source Control pane:** If you open the Source Control pane you see a list of modified, added, and deleted files available to stage and commit to the branch you're currently on.

>> **Extensions pane:** If you open the Extensions pane, you can install any useful extensions for your project, I recommend installing the GitHub Pull Request extension because then you can automatically track, check out, and update pull requests for your repository.

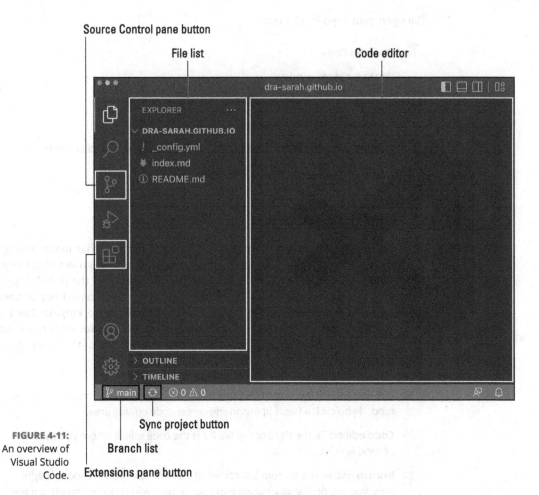

Source Control pane button

File list

Code editor

dra-sarah.github.io

EXPLORER

DRA-SARAH.GITHUB.IO
! _config.yml
index.md
README.md

> OUTLINE
> TIMELINE

main ⊗ 0 ⚠ 0

Sync project button

Branch list

FIGURE 4-11:
An overview of
Visual Studio
Code.

Extensions pane button

Finding Resources for GitHub Pages

Some amazing folks over at GitHub have dedicated all their time to supporting GitHub users in discovering and learning about all the features that GitHub offers. Beyond the static documentation, the GitHub Training Team offers guides, quickstarts, and feature updates directly from the GitHub pages docs site (`https://docs.github.com/pages`). You can even learn more through GitHub Skills (`https://github.com/skills/github-pages`).

GitHub Skills are a self-guided, automated tutorial that use repository templates to enable you to do the actions on GitHub.com and not just watching a video or reading a tutorial. You can find great ones for GitHub Pages, all GitHub fundamentals, Markdown, HTML, and even running your own Open Source Community. Head over to `https://skills.github.com/`, authenticate with your GitHub credentials, to try building some of these skills.

Chapter **5**

Creating a Website with GitHub Pages

I n this chapter, you can find effective strategies for reorienting yourself with an existing project, as well as specifics on building a website with GitHub Pages. The examples shown in this chapter assume that you have an existing GitHub Pages repository. If you don't and want guidance in setting one up, see Chapter 4.

You can follow along in this chapter to build a simple website on GitHub Pages or jump around to the various sections to find out how to do a specific task.

Jumping into an Existing GitHub Project

Whether you're revisiting a project that you started yesterday, one you worked on last year, or finding a new project that you've never worked on, there are quick and easy ways to get oriented with a GitHub project. In this section, you see examples of reorienting with the GitHub Pages website repo. (If you still need to set one up, see Chapter 4.)

To get started, make sure that you've opened your browser to GitHub.com and have signed in. If you need to create a GitHub.com account, you can read about how to do so in Chapter 1.

Accessing the GitHub.com repo

On the left side of the GitHub.com home page is the list of repositories that you have recently opened, contributed to, or created. Directly above the list is a search bar for searching repositories. This search bar becomes more useful as you interact with more GitHub repositories because the list of repositories can grow large, especially if you belong to a big organization.

At the top of the home page is another search bar that you can use to search for repositories, project boards, and teams (a feature of organizations beyond the scope of this book). By default, this top search bar is scoped to your current context. If you're on GitHub's home page, it searches all of GitHub. If you navigate to a repository on GitHub.com, the top search bar searches within that repository.

The search bar always gives you the option to search all of GitHub.com no matter where you are. Figure 5-1 shows the three ways to find a specific repository you may be looking for and two places to find new repositories.

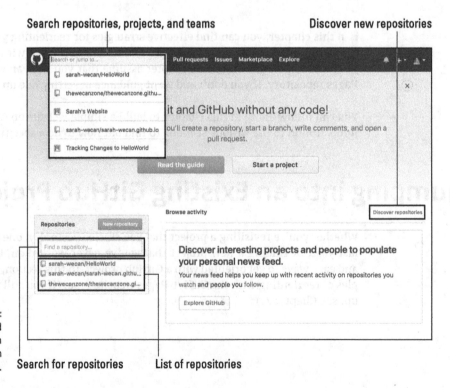

FIGURE 5-1: Places to find GitHub repos on the GitHub.com home page.

After you find the repository you want to start collaborating with, click it. The repository's home page appears.

TIP

This chapter gives specific examples about contributing to the GitHub Pages website repo that you're the owner of, so if you want to follow specifically for that, choose the repo titled *your-username*.github.io. The repo I use in this example is dra-sarah.github.io.

Verifying your permissions for the repo

If you're the owner or admin of a repository, you see a Settings tab at the top of the repository's home page. You have complete control over the repository, including the ability to

>> Invite collaborators.

>> Change the visibility of a repository from public to private or from private to public.

>> Delete the repository.

>> Archive the repository.

If you don't see the Settings tab, you're not the owner/admin, but you may have write permissions. To determine whether you do, you can attempt to make a change to a file by navigating to the file and clicking the pencil icon. If you're able to make a change and are presented with a Commit Changes box similar to the one shown in Figure 5-2, you have write permissions and were added as a collaborator for the project.

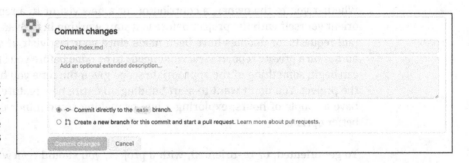

FIGURE 5-2:
The Commit Changes box when you have write permissions on a repo.

If you attempt to make a change but see the warning and commit box shown in Figure 5-3, you do not have write permissions and are not a collaborator. You can still create issues and propose file changes to the repo, but you have to get approval from someone with more permissions than you. If you're interested in how to contribute to projects that you don't have write access to, see Chapter 6.

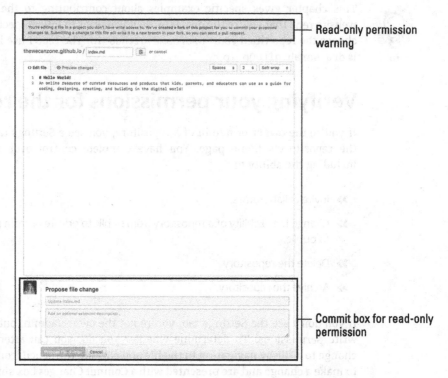

Read-only permission warning

Commit box for read-only permission

Orienting yourself with the project

Whether you're the owner, a contributor, or a new visitor to a repo, you need to orient yourself with the project before you start working in case additional issues, pull requests, or updates have been made since your last visit. If you're the only author on a private repo, review what needs to be done before you start so that you can begin something of the appropriate scope given the time you have to work on the project. You don't want to start building an entire new feature set if you only have a couple of hours; exploring a bug or making some minor updates may be better options.

To get oriented, or re-oriented, with a project, you should review four places on the repo:

>> **Read the README.md, CONTRIBUTING.md, and CODE_OF_CONDUCT.md files.** Unless you're working on your own project, you should always read through these three files at least to make sure that you understand how to

set up the project on your computer, contribute effectively to the project, and interact with other contributors. Not every project have all these files, but if they exist, you should read through them to ensure that you're a positive member of the project's community. You can find a good example of these documents on the GitHub Desktop open source project at https://github.com/desktop/desktop.

>> **Survey project boards.** If the project uses project boards, look at those boards. Click the Projects tab at the top of the repo and choose a project board. Each project board usually has a column of things in progress or that need to be done. If the project board has automation, then any changes to issues or pull requests (including new ones) appear on the project board.

>> **Read through issues.** Especially on active projects, folks are likely to have opened new issues. Click the Issues tab at the top of the repo to see the list of open issues. You can sort the issues by most recently updated to see whether folks have commented on existing issues. The default sorting is by newest (meaning the issues that were most recently opened). If you're the repository owner, triage any new or updated issues. *Triaging* is when you sort and order items. (See the sidebar called "Triaging issues" for more information about how and why to triage.)

>> **Review pull requests.** Both passively and actively reviewing pull requests is a good idea before starting to work on a project because they represent the new, removed, or changed code that aren't yet a part of the main branch. *Passively* reviewing pull requests means to read through the most recently opened and modified pull requests and see what kinds of contributions to the project are in the pipeline. This review helps to make sure that you don't start working on something that someone else is either already working on or that may break or contradict something that someone else is already working on. If you detect a problem, you can add a comment to the discussion on the pull request.

You can also *actively* review pull requests if you have the proper permissions for the project (and, more importantly, if you're confident that you can evaluate whether the changes should be merged). You can start the review process by opening a pull request, clicking the Files Changed tab, and then clicking the Review Changes button at the top right of the pull request, shown in Figure 5-4. In the example for this book, no pull requests are open for the website, but I've closed one (the initial website template that I added).

>> **Review discussions.** When you see a Discussion tab at the top of the menu bar, Discussions have been enabled for this repository. Clicking it takes you to a typical discussion board with categories, a Code of Conduct, and other useful information. You can see a good example of a GitHub Dicsussion board on the Sentry JavaScript repository (https://github.com/getsentry/sentry-javascript/discussions).

TRIAGING ISSUES

Just like with an emergency room full of patients, triaging issues is the process of sorting and classifying issues on your repo. Okay, triaging issues may not be just like an emergency room because the issues are hopefully not life-threatening, but they're often critical for your project. If you're the owner of a repo on GitHub, I recommend spending the first 30 minutes of your day on a project triaging anything that may have come in since the last time you visited the repo. In the triage process, you should include at least

- **Apply labels.** Issues and pull requests can have labels associated with them. A few must-have labels are

  - *bug:* An issue that reports something that appears to be broken and should be addressed quickly

  - *good first issue:* An issue that a new contributor could start with

  - *help wanted:* An issue where the code owner is specifically asking for help from the community

  - *needs investigation:* An issue with questions that need to be answered before a solution is known

  - *feature/enhancement:* An issue that requests a new feature or change to the project

  Some labels are added to a repo by default, including bug, duplicate, enhancement, good first issue, help wanted, invalid, question, and wontfix. You can delete labels you don't want to use and add other labels that you need repository needs.

- **Respond to comments.** Some issues may have new comments since the last time you were on the repo. For example, if someone opened an issue and you applied the label needs investigation and asked for additional information from the person who opened the issue, you may want to check to see whether they provided you with any additional information first. Issues should remain active. If an issue goes stale (as in folks stop commenting or making progress on it), it should probably just be closed.

- **Close stale issues.** Any issues that have gone stale should be closed. One example of when closing an issue is appropriate is when it's been a few weeks since the person who opened the issue last provided information or you know for a fact that you don't want to add a requested feature. Always comment on the issue before closing and let folks know why it was closed. These comments are also useful for your future self to remember why you closed an issue.

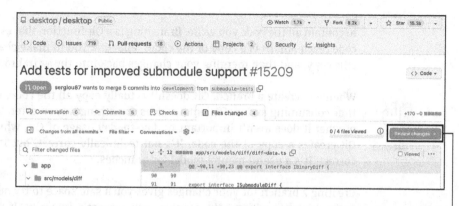

Start a pull request review

FIGURE 5-4:
The place to start
a review of a
pull request.

TIP

Chapter 8 covers the pull request review experience. However, if you want to have an interactive experience where you get to review a pull request on your own project before doing it on someone else's project, you can visit https://github.com/skills/review-pull-requests, click Use This Template, and follow the course described in the README written by the great folks at GitHub. You can also practice reviewing a pull request on a repo designed specifically for this book, which you can find at https://github.com/thewecanzone/GitHubForDummiesReaders/pull/2.

Preparing Your Contribution

After you orient yourself with your project, you need to decide what you're going to work on for your contribution. For the example in this book, I chose issue #2 in the Todo column of the project board: Change the title and tagline.

TIP

Other good candidates are any issues that were opened with the label bug, help wanted, or good first issue because these issues are typically urgent for owners or good entry points for contributors.

This section assumes you already have a repo cloned onto your machine. The examples in this section use GitHub Desktop and VS Code to resolve the issues. If you need guidance in cloning a repo or getting the project set up in GitHub Desktop or VS Code, see Chapter 4.

Creating a branch for your contribution

Before you modify or add code to any project, whether a private solo project you own or a large open source project, I highly recommend creating a specific branch to contain all the code you write. Branching is a Git function that essentially copies code (each *branch* is a copy of the code), allowing you to make changes on a specific copy, and then merging your changes back into the main branch.

REMEMBER

When you create a branch, Git doesn't actually copy all the code, which would be time consuming and inefficient. Git does something way smarter, but the specifics of what it does aren't important to day-to-day usage of Git, which is why I say Git creates a copy of the code. It's not technically correct, but it's conceptually correct. It's a useful mental model for branches.

Creating a branch for your changes gives you a safe space to try out solutions and make mistakes without threatening your project's integrity. If your solution is taking you down a wrong path, you can simply delete the branch, create a new one, and start over. Creating a branch helps your main branch remain in an always working state because you only merge your code into the main branch when you're confident it doesn't introduce any new problems and/or accurately solves the problem you were targeting.

Different projects and people use different nomenclatures for their branches. In this book, I use a short description of the code modifications to name branches. Another branch naming convention is to put the initials of the developer in front of each branch before a short description. Before creating a branch on a public repo, check out how other contributors have been naming their branches or see whether the CONTRIBUTING.md file contains any guidelines.

REMEMBER

This book uses GitHub Desktop to manage projects on your local machine and VS Code as the primary text editor. If you need to set up these two applications, see Chapter 4. You could also use VS Code directly, but I want to take this opportunity to show two useful dev tools.

To create the branch for your contribution, follow these steps:

1. **Verify the repo you cloned.**

   Open the GitHub Desktop application and verify that the top left drop-down menu has your website repo selected. Figure 5-5 shows the repo correctly selected.

2. **Start a new branch.**

   Click the branch drop-down list in the top center of GitHub Desktop and click the New Branch button (see Figure 5-5). When you click the New Branch button, a dialog box appears.

3. **Give your branch a name.**

   Type a name for your branch. I use new-title-and-tagline because that is the change I plan to make.

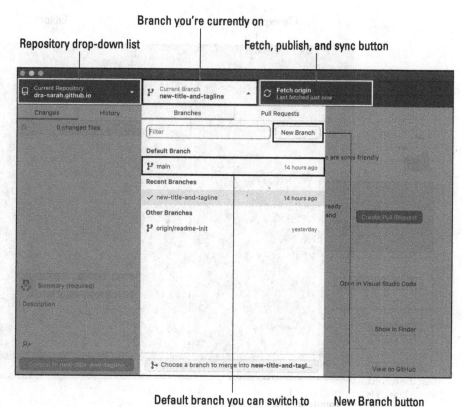

Repository drop-down list

Branch you're currently on

Fetch, publish, and sync button

Default branch you can switch to

New Branch button

FIGURE 5-5:
The button to create a new branch in GitHub Desktop for a specific repo.

4. **Click Create Branch.**

   The dialog box closes, and you switch to the new branch. Figure 5-6 shows the current branch as new-title-and-tagline, but the main branch is still available within the drop-down menu. The button at the top right of GitHub Desktop also changes from Fetch Origin to Publish Branch.

5. **Publish your branch.**

TIP

   Before making any changes to your branch, I recommend publishing your branch to GitHub.com. That way, as you start to modify or add code and push it to your branch, you won't lose any of your work. Click the Publish Branch button on the top right of GitHub Desktop. When the branch has been successfully published, this button goes back to the Fetch Origin button.

6. **Open your project in VS Code.**

   Open the VS Code application and verify that your project folder is open. A tree-view of your project appears in the lefthand Explorer pane. Notice that the branch selector at the bottom of the VS Code application window now shows the branch name. Figure 5-6 shows the correct branch checked out in VS Code.

File explorer pane   Editor

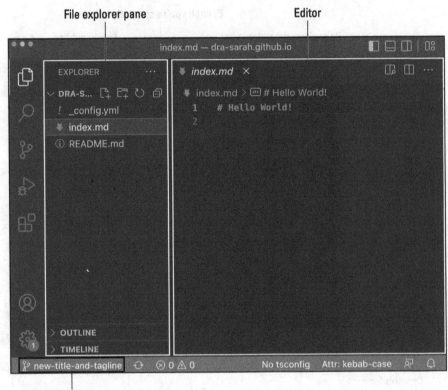

FIGURE 5-6:
The branch
selector in VS
Code showing
that the correct
branch is
checked out.

Currently checked out branch

TIP

If you click the branch selector in VS Code and change the branch, the branch in GitHub Desktop correctly changes to the same branch, and vice versa!

Confirming your branch is published

Before writing code that you want to keep, confirm that you correctly published your branch and are able to push to the branch. If you're working on the same project, on the same computer, then you most likely are still properly set up. But if you're contributing to a project for the first time or you have just set up a new computer, you should consider confirming you're able to contribute.

This section shows you examples of contributing to the GitHub website repo https://github.com/dra-sarah/dra-sarah.github.io using GitHub Desktop and VS Code. You should have your *username*.github.io repository cloned.

To get started, follow these six steps:

1. **Click a file in the file tree in VS Code.**

The file opens in the editor. Figure 5-6 shows index.md open.

2. **Modify the file.**

Don't make a lot of changes at this stage because the goal is to confirm you're able to push changes to the branch that you published.

3. **Save the file.**

Choose File ⇨ Save to save your changes. When you click Save, the color of the file in the file tree changes, and the Source Control button changes to indicate that a change occurred (see Figure 5-7).

Source control button showing changes ready to be staged/committed

A list of modified, added, or removed files   The editor showing a modified file

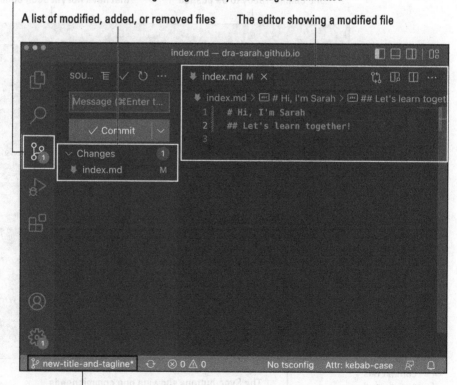

FIGURE 5-7:
VS Code after a modification has been made to the index.md file, but the change hasn't been committed.

The branch with an * showing there are changes that have not been committed or pushed yet

4. **Open the Source Control pane by clicking the Source Control button.**

You see the changes in the changes list.

The first time you're using VS Code to stage, commit, and push changes to a GitHub repo, if you try to commit changes that haven't been staged yet, you're asked whether you want to set all unstaged changes to be staged when you click the Commit button. For more information, you can check out the VS Code docs on Git support at `https://code.visualstudio.com/docs/editor/versioncontrol#_git-support`.

5. **Stage and commit the changes.**

Write a commit message. Then click Commit. Figure 5-8 shows how VS Code represents code that has been committed, but not yet pushed.

The Source control pane showing one commit ready to be pushed

The editor showing that there are no changes that have not yet been committed

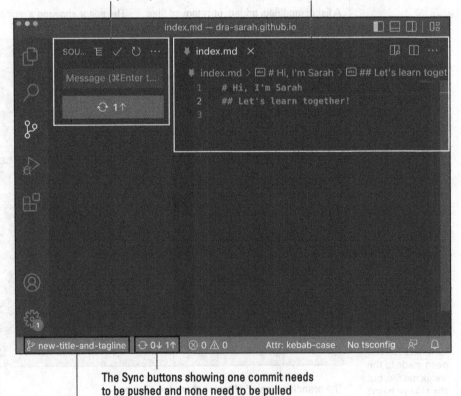

FIGURE 5-8:
VS Code after a commit has been made and before the local repository has been pushed.

The Sync buttons showing one commit needs to be pushed and none need to be pulled

The branch showing there are no uncommitted changes

6. Push and merge the changes.

The Sync button that replaces the Commit button in Step 5, or the bottom bar to the right of the branch name updates to show one commit is ready to be pushed to the remote branch. Click the Sync button to push your changes to GitHub. Then open GitHub.com and go to the repository. In this example, the repository is `https://github.com/dra-sarah/dra-sarah.github.io`. You now see a suggestion to open a new pull request (see Figure 5-9). If you don't, you can click the Pull Request tab, click Create New Pull Request, and create the pull request.

Create and merge the pull request. Chapter 3 gives a brief introduction to how to merge a pull request, and Chapter 8 gives an in-depth view of what you can do with pull requests if you need additional guidance.

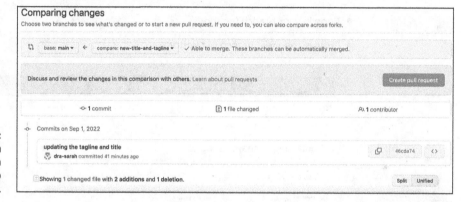

FIGURE 5-9: GitHub.com indicates when a branch is ready to be merged.

RESOLVING SIMPLE MERGE CONFLICTS

Sometimes your branch can get into a state where GitHub.com can't determine how to properly merge the code. This conflict can sometimes happen when a complex sequence of events happens. First, you create a branch from the main branch. As you make changes to your branch, someone else makes changes to the main branch. You then stage, commit, and push your changes to your branch, without pulling the changes made on the main branch into your branch first. The changes that you made may conflict with the changes that were made on the main branch — for example, you changed the same line, but started from a different place. For example, say that when you create your branch, the line is

```
Hello World!!!
```

and you changed it to

(continued)

(continued)

```
Sarah's Website
```

But someone else changed the line on the main branch to

```
Hello Friend!
```

When you create the pull request, GitHub gets confused. The commit shows that you changed code from Hello World!! to Sarah's Website, but the code that you're trying to merge says Hello Friend!, so it's unclear what should actually change. Rather than making a change for you that you don't intend to make, GitHub asks you what you want to do. When you create the pull request, instead of a button allowing you to merge the pull request, you're prompted with a button to resolve the conflicts. Clicking this button takes you to a new page that presents you with the line of code with two options (see the figure):

- Code in your branch (on top)
- Code in the main branch (on bottom)

You can edit the code by deleting the delimiter lines and choosing the line you want to keep. In this example, the final code looks like

```
# Sarah's Website
```

Then, you can click Mark as Resolved and commit the changes to your branch.

Sometimes, pull requests can create merge conflicts. See the sidebar "Resolving simple merge conflicts" for insight on how to resolve them.

Building Your Personal Website

This section is following the example website repo that is created in Chapter 4 to show some basics on how to build a website. However, the tips shown in this section can be used for any website you build using GitHub Pages.

Modifying the title and tagline

To modify the title and tagline of your website, open the _config.yml file in VS Code. Add two lines above the only line that is in the file, indicating the theme. Your code should look like this:

```
title: <Your Name>
description: <Your Description>
theme: jekyll-theme-cayman
```

Adding sections to your website

Adding sections to your website is made easy with Jekyll and GitHub Pages because you can use Markdown and HTML. First, include some social media usernames. Open your _config.yml file and add as many social media usernames as you want. In this example, I add a Twitter and GitHub username:

```
title: Your Name
description: Your Description
twitter_username: Your Twitter username
github_username: Your GitHub username
theme: jekyll-theme-cayman
```

Then, open the index.md file and change the code to include sections. For example, my code looks like this:

```
# My Projects
Here is a list of projects that I am working on:
# My Interests
I'm interested in teaching novice coders about computer science!
# My Blog
```

```
I'm really excited to blog my journey on GitHub.com.
# Get in Touch
<ul>
<li><a href="https://twitter.com/{{ site.twitter_username
    }}">Twitter</a></li>
<li><a href="https://github.com/{{ site.github_username
    }}">GitHub</a></li>
</ul>
```

Save, stage, commit, and push your changes to your branch and then create and merge the pull request into the main branch. After a couple of minutes, refresh the website page to see your changes.

Creating a blog

Having a blog on your site is a great way for you to share your GitHub journey with others. As you start to discover and create, you can share what you learn and build with a community of people with similar interests. This section guides you through creating the blog posts and linking them from your index.md file.

First, create a new folder called _layouts and create a file within that folder called post.html. The _layout/post.html file should contain the following code to create a blog-style:

```
layout: default
<h1>{{ page.title }}</h1><p>{{ page.date | date_to_string }} -
    {{ page.author }}</p>
{{ content }}
```

WARNING

Make sure that you correctly name the folder _layouts. Jekyll searches for the _layouts folder for any custom layouts. Otherwise, it uses the defaults for that theme. You can change the name of the specific layout — for example, post.md — but it should match the layout metadata, as shown in the following code snippet.

Using layouts and specific naming, Jekyll can extrapolate the title, date, and author to display that both on the blog post page and on the home page.

Then, create a new folder called _posts and a file inside with the date and a title. Typically, blog posts are made with YEAR-MONTH-DAY-TITLE.md. For example, this code is in a file called _posts/2019-01-01-new-year.md:

```
---
layout: post
```

```
author: sguthals
---
Write your blog post here.
```

Finally, in your `index.md` file, add the following code below the My Blog section:

```
<ul>
{% for post in site.posts %}
<li>
<a href="{{ post.url }}">{{ post.title }}</a>
</li>
{% endfor %}
</ul>
```

Save, stage, commit, and push your changes to your branch and then create and merge the pull request into the `main` branch. After a couple of minutes, refresh the website page to see your changes.

Linking project repos

You can link GitHub project repos to your website in the same way you link social media, described in the section "Adding sections to your website," earlier in this chapter. Putting a link directly to a repo can be efficient. However, you can also create web pages for project repos as well, as described in Chapter 4.

Open the `index.md` file and add the following code, replacing the URLs with links to projects you're the author of. This code shows linking to a project repo web page and directly to a project repo:

```
<ul>
<li><a href="https://sarah-wecan.github.io/HelloWorld/">Hello
    World Project</a></li>
<li><a href="https://github.com/thewecanzone/GitHubForDummies
    Readers">GitHub For Dummies Repo</a></li>
</ul>
```

Out of scope for this book are a vast number of ways you can customize your GitHub Pages website. Jekyll and GitHub come together to offer a unique experience that requires some coding, but handles a lot of the setup. To find out how to do something specific, start by visiting GitHub Help at `https://docs.github.com/pages/setting-up-a-github-pages-site-with-jekyll`.

3
Contributing to Your First Project

Fork your first GitHub repository so that you can contribute your own code.

Get unstuck when you've cloned and changed code before forking.

Create effective commit messages to communicate the changes you've made.

Create a pull request to start the process of your code being merged in.

Explore effective pull request workflows.

Review a pull request.

Chapter **6**

Forking GitHub Repositories

More than likely, you will want to work on some repositories where you are not the owner or collaborator. In situations where you aren't the owner or collaborator, you will have to fork the repo if you want to do anything other than browse the files.

In this chapter, I explain what forking is, show you how to fork a repository, and compare forking to cloning and duplicating. I also discuss contributing code via a fork. This chapter also demonstrates how to get the code you want to contribute into a fork if you've already made some changes to a clone before forking.

Introducing Forking

A *fork* of a repository is essentially just a copy of the repository. In the spirit of open source, forking is a way to share with and learn from other developers. Developers can have many motivations for forking a repository, but three of the most common reasons are to

>> Contribute to someone else's project

>> Use someone else's project as a starting point

>> Experiment with someone else's code without making changes to their project

If you aren't the owner or a collaborator on a project, you'll have to fork the repo if you want to do anything other browse the files. If you plan on making any changes to a repo where you aren't an author or collaborator, fork the repo first so that you're in the correct state as you start to explore and modify the code. Don't worry, though, if you forget to fork — see the section "Getting unstuck when cloning without forking," later in this chapter, for help.

Prior to GitHub, a fork of an open source project tended to have a negative connotation. It wasn't just a copy of the source code, but a split in the community. A fork implies a fork in the road, where one group takes the project in a new direction. For example, Joomla is a fork of the Mambo project by a group of people who felt like a company had too much control. In practice, forks tend to be good for the overall ecosystem because they introduce new ideas. In some cases, the best ideas in the fork make their way back into the original project, such as when EGCS, which forked from GCC, had its changes merged back into the GCC project. On GitHub, forks tend to be more like short-lived branches that are either merged back into the main code or deleted.

Cloning, Forking, and Duplicating

When you *clone* a GitHub repo, you're creating a local copy of the project on your computer. Forking a GitHub repo creates a copy of the repo on your GitHub.com account, and from there, you can clone the repo. A link between the original repo and the one you forked remains, allowing you to pull changes made on the original repo into your copy and push changes that you make on your copy to the original copy.

Duplicating a repo is when you make a copy that no longer has a link to the original copy. Duplication isn't a usual part of an open source workflow because it makes it more difficult to push changes back into the original repository. Even so, duplicating a repo can be useful sometimes, such as when the original project is no longer active and you plan to keep the project alive with your fork.

Cloning a Repository

Any public repo can be cloned, and you can run the code on your computer and make changes to the code. But you won't be able to push those changes back to the remote repo if you don't have push permissions to the repo. Chapter 4 describes how to clone a repository in GitHub Desktop when you're the owner. The process is the same whether you're an owner, collaborator, or visitor.

Before you clone a repository, you should verify that you're able to push changes to it. The easiest way to verify this ability is to go to the repository home page. If you see a Settings tab on the right side of the home page, then you likely have push rights. If you don't, you likely have to fork the repo first. Alternatively, you can try to edit a file on GitHub.com, and if you get the message shown in Figure 6-1, then you don't have permission to directly contribute to the project.

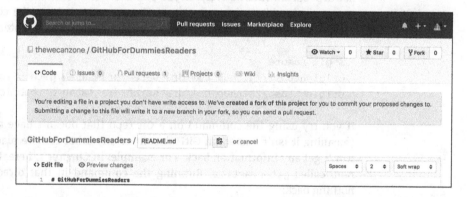

FIGURE 6-1: GitHub.com error message if you don't have edit permissions on a repo.

WARNING

If you do end up cloning a repository where you don't have permission to contribute to it, you can end up making changes and not being able to push them. The section "Getting unstuck when cloning without forking," later in this chapter, gives steps for how to get out of this state.

After you have a repository cloned on your local computer, you can view and modify metadata about it in your terminal. Open the terminal and go to a directory where you have a GitHub repository. If you need an example, clone https://github.com/thewecanzone/GitHubForDummiesReaders by typing

```
$ git clone https://github.com/thewecanzone/
   GitHubForDummiesReaders
Cloning into 'GitHubForDummiesReaders'...
remote: Enumerating objects: 15, done.
remote: Counting objects: 100% (15/15), done.
```

```
remote: Compressing objects: 100% (15/15), done.
remote: Total 15 (delta 4), reused 0 (delta 0), pack-reused 0
Unpacking objects: 100% (15/15), done.
$ cd GitHubForDummiesReaders
```

You can verify where the remote/target repo is with the following command:

```
$ git remote -v
originhttps://github.com/thewecanzone/GitHubForDummiesReaders.
    git (fetch)
originhttps://github.com/thewecanzone/GitHubForDummiesReaders.
    git (push)
```

If you cloned the same repo as I did, you see the exact same origin URLs for fetch and push. You should see that the remote repo is one owned by thewecanzone and not dra-sarah. Alternatively, if you run the same command on a repo that you own, you should see your username. For example, if I run the command in the directory where I cloned my website repo that I created in Chapter 4, I would see

```
$ git remote -v
originhttps://github.com/dra-sarah/dra-sarah.github.io.git (fetch)
originhttps://github.com/dra-sarah/dra-sarah.github.io.git (push)
```

If you try using the command on a Git repo that doesn't have a remote origin (meaning it isn't hosted on GitHub.com or any other remote place), you simply won't get any information back. For example, in Chapter 1, I created a simple Git repo called git-practice. Running the command in that directory gives you nothing back:

```
$ git remote -v
```

Forking a Repository

The goal of open source is to encourage collaboration among software developers around the world, so being able to contribute code to repositories where you aren't the owner or an explicit collaborator is an important part of the GitHub workflow and mission. To become a collaborator of an open source project, you can reach out to the owner of the repo and request to be a collaborator. However, if the owner doesn't know who you are, they probably won't add you as a collaborator because that would give you push rights to the repository. You'll have to gain their trust first.

REMEMBER

You don't need the owner's permission to fork their repository. You can make your contributions and share them with the owner to show how you can be an asset to the project.

To fork a repo, go to the repo home page and click the Fork button at the top right. If you'd like, you can use https://github.com/thewecanzone/GitHubForDummiesReaders to practice forking and contributing to a public repo.

After you click the Fork button, the web page refreshes, and you see a slightly modified version of the repo, as shown in Figure 6-2. At the top of the repo, you see that the repo is attached to your account, but it still has a reference to the original repo.

Crumb trail linking the forked repo with the upstream repo

FIGURE 6-2:
A forked repo on
GitHub.com.

WARNING

If you're part of multiple GitHub organizations, you're asked to choose which organization you want to fork the repository to after you click the Fork button and before the web page refreshes.

After you have your own, forked version of the repo, you can clone it on your local machine to start making changes. Chapter 7 goes over writing code and creating commits, which is the same process whether you're on a forked repo or a regular repo. If you clone the repo using GitHub Desktop, your local Git repository knows about your forked version (remote origin) and the original repo (remote upstream).

TIP

The concept of a *remote* can be confusing to those new to distributed version control systems like Git. When you clone a GitHub repository, you have a full copy of the repository on your local machine. You may be tempted to think the copy of the repository on GitHub is the canonical copy. However, there is no concept of canonical in Git. The *canonical copy* is whatever the people working on the project decide it is by consensus. Git does have the concept of a *remote*, which is a pointer to a copy of the same Git repository hosted elsewhere. Typically, a remote is a URL to a Git-hosting platform like GitHub, but it's possible to be a path to a directory with a copy of the repository. When you clone a repository, Git adds a remote named origin with the location (usually a URL) from where you cloned it. But it's possible to add multiple remotes to a Git repository to indicate other locations where you may want to push and pull changes from. For example, if you clone a fork of a repository, you may want to have a remote named upstream that points to the original repository.

If you clone the repo using the command line, you may want to set the upstream remote, which I explain in the section "Getting unstuck when cloning without forking," later in this chapter. You can see both the forked remote origin and original remote upstream if you run the git remote -v command in the directory where you cloned the repo:

```
$ git remote -v
originhttps://github.com/dra-sarah/GitHubForDummiesReaders.git
  (fetch)
originhttps://github.com/dra-sarah/GitHubForDummiesReaders.git
  (push)
upstreamhttps://github.com/thewecanzone/GitHubForDummiesReaders.
  git (fetch)
upstreamhttps://github.com/thewecanzone/GitHubForDummiesReaders.
  git (push)
```

The *origin*, which is your fork of the repository where you typically fetch/pull changes from and push changes to, has your username (dra-sarah in this example). The *upstream*, which is where the original code is located, and where you eventually want to contribute the code you write back to, has the original author's username (thewecanzone in this example). While you can push changes to and pull changes from any remote (origin or upstream), it is good practice to work on your fork of the repositoy, represented typically as *origin*.

Fetching changes from upstream

Having the upstream repo linked to your forked repo is important. As you start making changes, you want to be able to fetch/pull any changes that are being made on the original code into your code to make sure that you have the most up-to-date version.

For example, suppose that you forked and cloned a website project a week ago with plans to change the website's About page. While you were working on those changes, someone else made a change to the About page. Their changes may conflict with your changes, or they may introduce something new that you want to use in your changes. Pulling those changes into your local repository before you submit your changes back to the original repository makes sense. It reduces the chance that your changes conflict with the changes others are making to the About page and makes it more likely the owner can accept them.

If you find yourself in a situation where you need to get the change from the upstream, original repo, you can go to the directory where your forked repo is and type

```
$ git fetch upstream
remote: Enumerating objects: 5, done.
remote: Counting objects: 100% (5/5), done.
remote: Compressing objects: 100% (3/3), done.
remote: Total 3 (delta 1), reused 0 (delta 0), pack-reused 0
Unpacking objects: 100% (3/3), done.
From https://github.com/thewecanzone/GitHubForDummiesReaders
8404f3b..e02a4d2 master -> upstream/master
$ git checkout -b new-branch
Switched to a new branch 'new-branch'
$ git merge upstream/master
Updating 8404f3b..e02a4d2
Fast-forward
README.md | 2 +-
1 file changed, 1 insertion(+), 1 deletion(-)
```

These three commands fetch the changes from the upstream repo, ensure that you're on your local, forked repo on a new branch, and then merges the changes from the upstream repo into your forked repo.

Contributing changes to upstream

After you make changes and publish them to a new branch in your forked repository, you're ready to suggest your changes to the original owner. If you go to the original, upstream repo on GitHub.com, your branch shows up on the home page, and GitHub asks whether you want to open a pull request to merge the changes with the original repo (see Figure 6-3).

On your forked repo on GitHub.com, your branch shows up, and GitHub asks whether you want to create a pull request for it (see Figure 6-4).

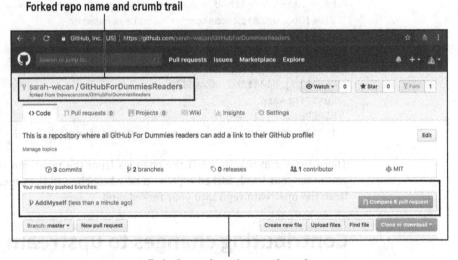

Upstream repo name

FIGURE 6-3:
Original,
upstream repo
detecting a new
branch from a
forked repo.

Upstream repo detecting branch from forked repo

Forked repo name and crumb trail

FIGURE 6-4:
Forked repo
detecting a
new branch.

Forked repo detecting new branch

Click the Compare & Pull Request button, and a pull request creation page gives you the option to request to merge your changes with the upstream repo or your forked repo (see Figure 6-5). Choose the upstream repo, add a comment, and click the Create Pull Request button.

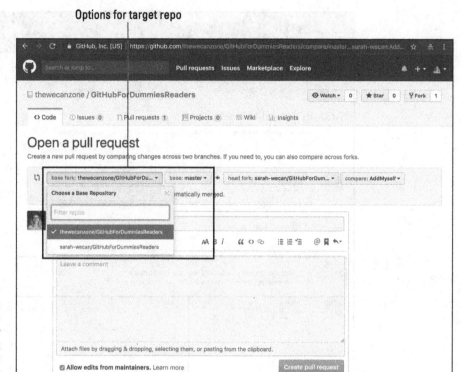

Options for target repo

FIGURE 6-5:
Forked repo
detecting a new
pull request
with the option
for upstream
or forked.

You see that the branch can be merged, but you have no way to personally merge the pull request because you aren't the owner of the target branch (upstream, original repo); only the owner (or a specified collaborator) has permission to merge code. Figure 6-6 shows the pull request on your repo without the option to merge.

As the owner of the upstream, original pull request, I can see the pull request and have the option to merge it (see Figure 6-7). If you're creating a pull request on this repo, I will continually merge pull requests so that I can keep an up-to-date table of all the *GitHub For Dummies* readers!

TIP

If you have a lot of changes that you want to add to your fork before requesting that they get merged into the upstream, original repo, then you can first create the pull request to target your forked repo instead of the upstream repo. This is a change in what is shown in Figure 6-5. When you're ready to merge your changes into upstream, you can create a new pull request to request the target of the merge be the upstream repo.

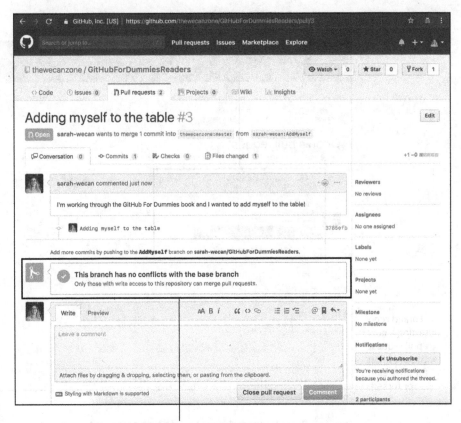

FIGURE 6-6:
Pull request
without the
option to merge.

Unmergeable pull request status tile

Getting unstuck when cloning without forking

One common problem people run into is they forget to fork a repository before they try to contribute to it. The following scenario describes one example of getting into this situation.

Here's the scenario: You clone a repository onto your local computer, modify the code, commit changes to main, and are ready to push your changes. But then you get a scary-looking error message. You may get the message in VS Code (see Figure 6-8), GitHub Desktop (see Figure 6-9), or in the terminal:

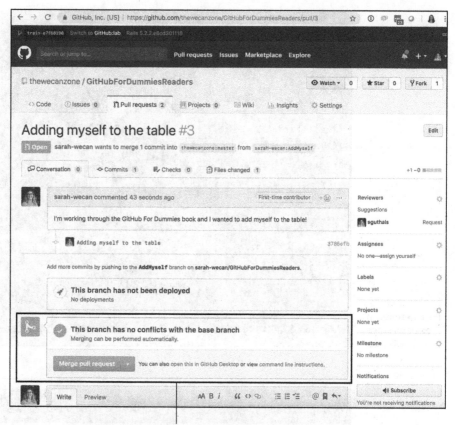

FIGURE 6-7:
Pull request with the option to merge.

Mergeable pull request status tile

```
$ git push origin main
remote: Permission to thewecanzone/GitHubForDummiesReaders.git
    denied to dra-sarah.
fatal: unable to access 'https://github.com/thewecanzone/
    GitHubForDummiesReaders.git/': The requested URL returned
    error: 403
```

The error message tells you that you don't have permission to push to this repository. You should have forked the repository first. You also made the mistake of committing directly to the development branch. As I recommend elsewhere in the book, it's a good practice to make all your changes in a temporary branch; this applies to any protected branches such as main or development.

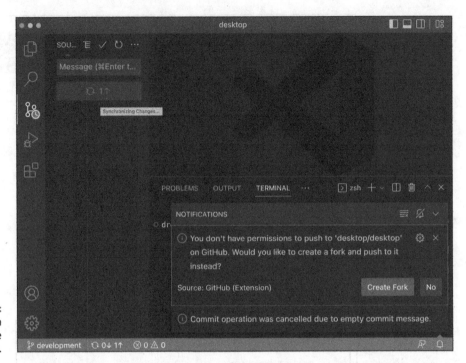

FIGURE 6-8:
Push permission
error message
in VS Code.

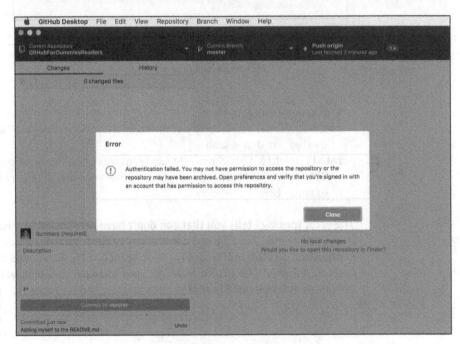

FIGURE 6-9:
Push permission
error message in
GitHub Desktop.

To fix this mistake, you need to move your changes to a new branch, fork the repo, change the remote URLs for your local repository to point to your fork, and push your changes. This process can get tricky, but if you don't follow the VS Code prompts, these steps can help you out of this predicament from the terminal:

1. **Migrate your changes to a new branch.**

   Right when you discover you're targeting the incorrect remote repository, you should move your changes to a new branch. You don't want to accidentally pull in changes from the upstream, original branch onto all the hard work you just finished. This step can get tricky, but luckily there's a Git alias to help. See the nearby sidebar "Creating a Git Alias" for help. After you have the `git migrate` alias, go to the directory where your repo is in your terminal and type

   ```
   $ git migrate new-branch
   Switched to a new branch 'new-branch'
   Branch 'main' set up to track remote branch 'main' from
     'origin'.
   Current branch new-branch is up to date.
   ```

   Confirm that the new branch has been created:

   ```
   $ git status
   On branch new-branch
   nothing to commit, working tree clean
   ```

   TIP

   You can also confirm that your commits are only in this new branch and no longer in the old branch by running a `log` command to compare the two branches:

   ```
   $ git log main..new-branch --oneline
   ```

   This lists the commits in new-branch that are not in main. The --oneline flag prints each commit on a single line, which is useful when you just need a summary of commits and not the full details.

2. **Set the upstream remote to be the original GitHubForDummiesReaders repo.**

   To add an upstream remote to your repo, go to the terminal and type

   ```
   $ git remote add upstream https://github.com/thewecanzone/
     GitHubForDummiesReaders.git
   ```

Confirm that the upstream remote was added correctly:

```
$ git remote -v
originhttps://github.com/thewecanzone/
    GitHubForDummiesReaders.git (fetch)
originhttps://github.com/thewecanzone/
    GitHubForDummiesReaders.git (push)
upstreamhttps://github.com/thewecanzone/
    GitHubForDummiesReaders.git (fetch)
upstreamhttps://github.com/thewecanzone/
    GitHubForDummiesReaders.git (push)
```

3. **Fork the repo.**

   Back on GitHub.com, go to the original repo and click Fork at the top right of the repo home page. The page refreshes, and you see your own version of the repo, referencing the original repo (refer to Figure 6-2).

4. **Set the origin remote to be your forked repo.**

   After you have your own fork of the repo, you can change your remote origin to be your version:

```
$ git remote set-url origin https://github.com/dra-sarah/
    GitHubForDummiesReaders.git
```

   You can also confirm that all your remote URLs are correctly set:

```
$ git remote -v
originhttps://github.com/dra-sarah/GitHubForDummiesReaders.
    git (fetch)
originhttps://github.com/dra-sarah/GitHubForDummiesReaders.
    git (push)
upstreamhttps://github.com/dra-sarah/GitHubForDummiesReaders.
    git (fetch)
upstreamhttps://github.com/dra-sarah/GitHubForDummiesReaders.
    git (push)
```

5. **Push your branch to your forked version.**

   You're now in the same state that you would be in had you forked the repo before cloning. Back in VS Code, you can publish your branch.

6. **Create a pull request.**

   Your forked repo detects a new branch and offers to have you create a pull request (refer to Figure 6-4).

CREATING A GIT ALIAS

A *Git alias* is an easy way to automate and extend Git commands. If you're doing a lot of Git commands on the terminal, creating Git aliases can make your software development more efficient. For example, in your terminal you can type

```
$ git config --global alias.st status
```

Now, instead of typing git status, you can type git st, and Git returns the current status of your repository. Getting rid of just four letters may seem a little silly, but it can end up making your Git command experience a lot more efficient over time.

A Git alias is a lot more powerful than just reducing the number of keys you have to press. You can read about a tricky scenario at https://haacked.com/archive/2015/06/29/git-migrate where you have to migrate the commits you've made on a branch to another branch. This migration is critical if you get stuck in the position where you've started working on a clone of a repository where you don't have write permissions, as I discuss in the section "Getting unstuck when cloning without forking," earlier in this chapter.

The Git alias to migrate commits from one branch to another is complex. It's a few complicated steps all rolled into one simple git migrate command. To make this command accessible for you to use when you get stuck on an unforked clone, follow these steps in your terminal:

```
$ open ~/.gitconfig
$
```

Your .gitconfig file opens in your default editor. Add the following code to the bottom of your .gitconfig file:

```
[alias]
migrate = "!f(){ CURRENT=$(git symbolic-ref --short HEAD); git
    checkout -b $1 && git branch --force $CURRENT ${3-$CURRENT@
    {u}} && git rebase --onto ${2-master} $CURRENT; }; f"
```

If your .gitconfig file already has an [alias] section, don't retype that line. Save and close the .gitconfig file.

Now you can use the git migrate command to migrate commits from one branch to another branch! This Git command has one required parameter and two optional parameters:

```
git migrate <new-branch-name> <target-branch> <commit-range>
```

(continued)

(continued)

The parameter `<new-branch-name>` is required. This branch is where you move the commits to. If you don't specify anything else, then the `migrate` command moves all commits from the `main` branch to this new branch.

The parameters `<target-branch>` and `<commit-range>` are optional. `<target-branch>` allows you to move commits from a branch other than the `main` branch to the `<new-branch-name>` that you specify in the first parameter. `<commit-range>` allows you to specify which commits you want to move over. This parameter can be useful if you accidentally made one commit on the wrong branch, and you just want to move that one commit over to `<new-branch-name>`.

Chapter **7**

Writing and Committing Code

I n this chapter, you write and commit code. The first part, writing code, is a very broad topic — too broad to be covered in this (or any single) book. The code I write in this chapter sets the stage for covering how to create good commits. Most of this chapter focuses on committing code. No matter what kind of code you write, the act of committing that code remains the same.

The code example I use throughout this chapter may seem contrived and overly simplistic. That's because it is contrived and simple. Don't let the simplicity, though, distract you because the information in this chapter also applies to large code bases.

Creating a Repository

A *commit* is the smallest unit of work with Git. It represents a small logical group of related changes to the repository. A commit additionally represents a snapshot in time — the state of the entire repository can be represented by referencing a single commit.

Before writing code, you need to create a local repository to store the code. In the following examples, I create a repository in a directory named best-example. Feel free to change best-example to a directory of your choice. Fortunately, this process is quick and painless:

1. **Open the terminal on your computer.**

   If you don't know how to do so, see Chapter 1 for guidance.

2. **Go to the directory where you want your project folder to be stored and type the following commands:**

   ```
   $ git init best-example
   $ cd best-example
   ```

   The first command creates an empty Git repository in the specified directory, best-example. Because the best-example directory doesn't already exist, Git creates it. The second command changes the current directory to this new directory.

TIP

Nearly every Git tutorial I've seen that covers initializing a Git repository does it in the current directory by calling git init with no parameters or git init . where the . represents the current directory. People can be forgiven for not realizing you can both create the repository directory and initialize it in one step by passing in the path to the new repository like I do here. In fact, you can combine both of these commands into a single command: git init best-example && cd best-example. This tip can help you gain the admiration and adulation of your less efficient peers!

Writing Code

After you're in a Git repository directory, you can start adding files. (If you aren't in a directory, see the previous section, "Creating a Repository" where I created the best-example directory.)

For this example, you create three files by typing the following code:

```
$ touch README.md
$ touch index.html
$ mkdir js
$ touch js/script.js
```

Note that one of the files you create is a README.md file. To find out why every repository should have a README.md file, see Chapter 10.

After running these commands, you have three files:

>> README.md

>> index.html

>> script.js

script.js is in a subdirectory named js. You guessed it — you're making a simple website!

You can flesh out the README.md file first. In this example, I use VS Code to open and edit the files in the current directory. (If you need any guidance setting up VS Code, see Chapter 2.)

TIP

Make sure you have installed VS Code on your PATH. If you need help, you can follow the getting started guides at https://code.visualstudio.com/docs/setup/setup-overview.

You can add some simple Markdown text to the README.md document. *Markdown* is language that offers a simple way to format and style your text. You can check out a guide on Markdown on the GitHub guides https://guides.github.com/features/mastering-markdown.

Open the README.md in the editor by clicking in the file tree in VS Code. Then add some Markdown relevant to your project. In this example, add the following text:

```
# The Best Example Ever
Which will be a part of the best commit ever.
```

Then add the following code to index.html.

```
<!doctype html>
<html lang="en">
  <head>
        <meta charset="utf-8">
        <title>It is the cod3z</title>
        <script src="js/script.js"></script>
  </head>
  <body>
        <h1>The Best Cod3z!</h1>
  </body>
</html>
```

This HTML file references `script.js`. Open `script.js` in VS Code and add the following code.

```
document.addEventListener(
        "DOMContentLoaded",
        function(event) {
        alert('The page is loaded and the script ran!')
        }
);
```

Make sure to save your changes to each file. Now test the code by opening `index.html` in your browser from the terminal.

```
$ open index.html
```

The page loads in your default browser, and the alert message, shown in Figure 7-1, appears.

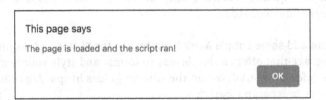

FIGURE 7-1:
An alert
message from my
running code.

Creating a Commit

This section assumes you have code that you've changed on your local computer and that the code is in a working state. If you need an example of working code, see the previous section in this chapter to get to this state.

After you have running code, you can commit it to the repository. To create a commit is a two-step process:

1. **Stage the changes you want to commit.**

2. **Create the commit with a commit message.**

Staging changes

Staging changes can be confusing to the Git beginner. In concept, it's similar to a staging environment for a website. Staging changes is an intermediate place where you can see the changes you're about to commit before you commit them.

Why would you want to stage changes before committing them? In Git, a commit should contain a group of related changes. In fact, Git encourages this setup.

Suppose that you've been working for a few hours and now have a large set of unrelated changes that aren't committed to the Git repository.

You may be tempted to just commit everything with some generic commit message like "A bunch of changes." In fact, an XKCD comic makes light of this phenomena at https://xkcd.com/1296.

Committing a bunch of unrelated changes is generally a bad idea. The commit history of a repository tells the story of how a project changes over time. Each commit should represent a distinct cohesive set of changes. This approach to commits isn't just about being fastidious and organized. Having a clean Git history has concrete benefits.

TIP

One benefit of a clean Git history is that a command like `git bisect` is way more useful when each commit is a logical unit of work. The `git bisect` command is an advanced command, and full coverage of what it does is beyond the scope of this book. In short, though, `git bisect` provides a way to conduct a binary search through your Git history to find the specific commit that introduces a particular behavior, such as a bug. If every commit contains a large group of unrelated changes, finding the specific commit that introduces a bug isn't as useful as it would be if every commit contains a single logical unit of change.

In the example for this chapter, I can probably stand to create two commits:

>> One that just contains the README.md file

>> Another that contains the index.html and script.js files

Because the index.html file references the script.js file, checking in one without the other doesn't make sense at this point.

Start by staging the README.md file:

```
$ git add README.md
```

The README.md file is added to the Git index. The Git index is the staging area for creating commits to the repository. You can check the status of the repository to see that the file has been added to the index:

```
$ git status
On branch main
No commits yet
Changes to be committed:
(use "git rm --cached <file>..." to unstage)
        new file: README.md
Untracked files:
(use "git add <file>..." to include in what will be committed)
        index.html
        js/
```

As you can see, the README.md file is staged for commit. Meanwhile, the index.html and js/ directory aren't yet tracked by this repository.

TIP

Why isn't script.js listed in the untracked files section? Git is taking a shortcut here. It notices that no files within the js/ directory are tracked, so it can simply list the directory rather than list every file in the directory. In a larger code base, you'll be glad Git isn't listing every file in every subdirectory.

Committing a file

After you stage changes (see preceding section), you can create a commit. In this example, I use the –m flag with the git commit command to specify a short commit message. The following commands demonstrate how to create a commit and specify the commit message in one step:

```
$ git commit -m "Add a descriptive README file"
[main (root-commit) 8436866] Add a descriptive README file
1 file changed, 3 insertions(+), 0 deletions(-)
create mode 100644 README.md
```

The file is committed. If you run the git status command again, you see that you still have untracked files. The git commit command commits only the changes that are staged.

Committing multiple files

After you commit the first file, you're ready to stage the rest of the files for a commit.

```
$ git add -A
$ git status
On branch main
Changes to be committed:
(use "git reset HEAD <file>..." to unstage)
        new file: index.html
        new file: js/scripts.js
```

The -A flag indicates that you want to add all changes in the working directory to the Git index. When you run the `git status` command, you can see that you've staged two files.

When the js directory is untracked, `git status` lists only the js directory and none of the files in the directory. Now that you're trying to stage the js directory, Git lists the file in the js directory. Why the discrepancy? Git doesn't actually track directories. It tracks only files. Therefore, when you add a directory to a Git repository, it needs to add each file to the index.

Sometimes you need to write a more detailed commit message. In this example, I didn't specify a commit message when I run the commit command because I plan to write a more detailed commit message:

```
$ git commit
```

If you don't specify a commit message using the -m flag, Git launches an editor to create a commit message. If you haven't configured an editor with Git, it uses the system default editor, typically VI or VIM.

There are legions of jokes about how difficult it is to exit VIM, so I won't rehash them all here. I'll simply take a moment of silence in remembrance for friends still stuck in the VIM editor.

For the record, to exit VIM, press the ESC key to exit the edit mode and type :wq to exit and save or :q! to exit without saving.

To change the default editor to something like VS Code, run the following command in the terminal:

```
gitconfig --global core.editor "code --wait"
```

The editor opens a temporary file named COMMIT_EDITMSG, which contains some instructions that are commented out:

```
# Please enter the commit message for your changes. Lines starting
```

```
# with '#' will be ignored, and an empty message aborts the commit.
#
# On branch main
#
# Initial commit
#
# Changes to be committed:
# new file: README.md
#
# Untracked files:
# index.html
# js/
#
```

You enter your commit message in the file that gets opened. You can write your message before all the comments or simply replace everything in the file with your own commit message.

In this case, I replaced everything in that file with

```
Add index.html and script.js
This adds index.html to the project. This file is the
default page when visiting the website.
This file references js/script.js, which is also added
in this commit.
```

After you save the commit message and close the file or editor, Git creates a commit with the message you wrote.

Writing a Good Commit Message

What should you write in a commit message? What makes a good commit message?

A Git commit should contain a logical and cohesive change or set of changes. The message should describe that change in clear terms so that anyone who reads the message later understands what changed in the commit.

REMEMBER

The audience for the commit message are current and future collaborators on the project. Those collaborators may include yourself in the future. Someday you may be tracking down a bug and want to understand why you made some change. You'll thank past-you for writing a well-written commit message that answers that question. So write clear commit messages and be nice to future-you.

If you find that you have trouble describing a commit, it may be that the commit contains too many changes. In writing code, well written functions do one thing and do it well. Similarly, a commit should represent one change to the system. The commit message describes the change and why it's being made.

A good commit message should also follow a specific structure. In general, a commit message has two parts:

>> The **summary** should be short (50 characters or less) and in the imperative present tense. For example, instead of writing "I added a method to Frobnicate widgets," write "Add method that Frobnicates widgets."

>> The **description** provides detailed explanatory text, if needed. Not every change requires explanatory text. For example, if you rename a function, a commit message with just a summary of "Rename Frobnicate to Bublecate" may suffice. You should wrap the description text at 72 characters. This ensures it looks good when displayed in the terminal as part of the output from the git log command.

By convention, a new line character separates the summary from the description.

Here's an example commit message in an open source project https://git.io/fhZ5a:

```
Avoid potential race condition
In theory, if "ClearFormCache" is called after we
check `contains` but before we execute the `return`
line, we could get an exception here.
If we're concerned about performance here, we could
consider switching to the ConcurrentDictionary.
```

There are a few conventions you can use within a Git commit message that Git will ignore, but GitHub will recognize. For example, you can specify that a commit resolves a specific issue with something like "fixes #123" where 123 is the issue number. When a commit with this pattern is pushed to GitHub, the issue number is linked to the issue. And when the branch that contains that commit is merged into the default branch of the repository (typically main), GitHub closes the referenced issue. That's pretty handy!

TIP

You can also use emojis in a commit message. For example, one pattern some teams use is to indicate that a commit contains only cosmetic changes by prefacing it with :art:. When that commit is rendered on GitHub.com, GitHub renders the emoji. You can see this in action in Figure 7-2, which shows a list of commits from the GitHub for Visual Studio open source project https://git.io/fhnDu.

The "Using GitHub Conventions in Commit Messages" section also outlines many helpful things you can include in your messages.

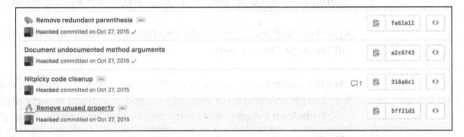

Committing Code with GitHub Desktop

Even though committing from the terminal is pretty straightforward, many people prefer to use a Graphical User Interface (GUI) application to commit code. Using a GUI has these benefits:

>> A GUI can provide guidance on conventions with commit messages, such as keeping the summary to 50 characters and separating it from the description by new lines. A GUI can simply present two fields: summary and description.

>> A GUI can provide support for GitHub specific conventions, such as the one where you can specify that a commit resolves an issue.

GitHub Desktop is a GUI created by GitHub that is great for committing code.

Tracking a repository in Desktop

Choose a repository that you have never opened in Desktop, but that you have locally on your computer. (See Chapter 2 if you haven't worked with Desktop yet.) If you need an example, use the best-example repository that you can create in the section "Creating a Repository," earlier in this chapter. When you launch Desktop, the best-example repository isn't listed in the list of repositories. Desktop doesn't scan your computer for Git repositories to manage. Instead, you have to tell Desktop about each repository you want to manage.

As expected, if you use Desktop to clone or create the repository, it's already tracking it. But sometimes you have a repository that you cloned or created outside of Desktop — for example, I created best-example using the terminal. Now you need to tell Desktop to track the repository you have chosen. Fortunately, this task is easy from the terminal.

The Desktop command line tool allows you to launch Desktop from your terminal, which allows you to easily integrate Desktop as much or as little as you want into your existing terminal-based Git workflow.

On Windows, you don't need to install the command line tool; it's done automatically. On the Mac, you have to take a separate step.

To install the command line tool on a Mac:

1. **Make sure Desktop is the active application and then, in the application menu bar, choose GitHub Desktop ⇨ Install Command Line Tool.**

2. **From the terminal, make sure that you're in the repository you want Desktop to track.**

   For this example, I'm in the best-example.

3. **Run the following command:**

   ```
   $ github .
   ```

   The . in the command represents the current directory. It could, instead, be a fully qualified path to a directory. GitHub Desktop launches (if it's not already running) and opens the specified directory. Because the current directory is already a Git repository, Desktop adds it to the list of repositories that it tracks. It then sets this repository as the current repository so that you can browse the repository's history, switch branches, and create commits, as shown in Figure 7-3.

If the current directory wasn't a repository, Desktop prompts you to create a Git repository in that directory. How convenient!

Publishing a repository in Desktop

To experience the full power of Desktop's integration with GitHub, you need to publish this repository to GitHub:

1. **Clicking the Publish repository button.**

   A dialog box to publish the repository appears (see Figure 7-4).

2. **Fill in the details and click the Publish Repository button.**

   The repository is created on your GitHub.com account.

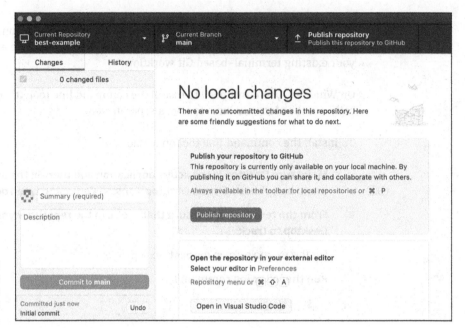

FIGURE 7-3:
GitHub Desktop
opened to the
best-example
repository.

FIGURE 7-4:
The Publish
dialog box used
to publish a
repository to
GitHub.com.

TIP

Desktop provides a keyboard shortcut to open the browser to the repository: ⌘-Shift-G (Ctrl+Shift+G on Windows).

If you want, you can create a few issues in the repository. (See Chapter 3 to find out how to create issues.) For this example, I created five issues:

>> Provide more details on the README.

- >> Mention in the README that this is a collaborative effort.
- >> Add a contribution section to README.
- >> Do not use an alert message.
- >> Set up website alerts.

You can also see these issues at this repo: `https://github.com/FakeHaacked/best-example/issues`.

Committing in Desktop

Desktop is used only for Git operations. To edit the files in the repository, you still need to use your editor of choice.

Make some changes so you have something to commit. In the example for this chapter, you can make some changes to `index.html` shown in bold.

```html
<!doctype html>
<html lang="en">
        <head>
                <meta charset="utf-8">
                <title>The Best Example</title>
                <script src="js/script.js"></script>
        </head>
        <body>
                <h1>The Best Cod3z!</h1>
                <div id="message"></div>
        </body>
</html>
```

Update `script.js` to populate the new DIV element, like civilized people would, rather than use an alert message. Changes are in bold:

```javascript
document.addEventListener(
        "DOMContentLoaded",
        function(event) {
        var message = document.getElementById('message')
        message.innerText = 'The script ran!'
        }
);
```

Switch back to Desktop and click the Changes tab, shown in Figure 7-5.

The left pane lists the set of changed files. If you select a file, you can see the

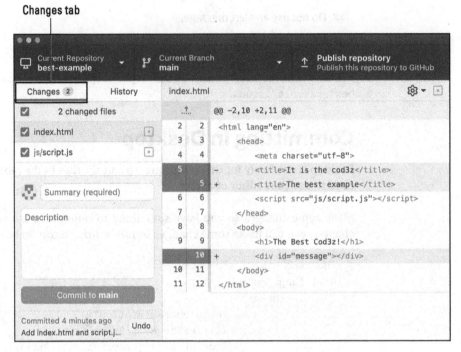

Changes tab

FIGURE 7-5:
The Changes
View showing the
uncommitted
changes.

Inside the diff view figure:

Current Repository
best-example

Current Branch
main

Publish repository
Publish this repository to GitHub

Changes 2   History   index.html

2 changed files

index.html

js/script.js

Summary (required)

Description

Commit to main

Committed 4 minutes ago   Undo
Add index.html and script.j...

```
@@ -2,10 +2,11 @@
 2   2   <html lang="en">
 3   3     <head>
 4   4       <meta charset="utf-8">
 5       -     <title>It is the cod3z</title>
     5   +     <title>The best example</title>
 6   6       <script src="js/script.js"></script>
 7   7     </head>
 8   8     <body>
 9   9       <h1>The Best Cod3z!</h1>
    10   +     <div id="message"></div>
 10  11     </body>
 11  12   </html>
```

specific changes to that file in the right pane, which is called the *diff view*.

You can commit all the changes as a single commit, but sometimes you might have unrelated changes. In this example, I have two unrelated changes:

>> The change to the title in index.html

>> The message changes to both index.html and script.js

Each of these changes should be in their own commit. How do you do that when index.html contains two unrelated changes? Fortunately, Desktop provides a nice way to commit a portion of the changes in a file. This process is known as a partial commit:

1. **Deselect all changes.**

   In the left pane, uncheck the check box next to the label 2 changed files to deselect all changes.

2. **Select index.html in the left pane.**

Click the filename in the left pane to display the changes for index.html.

3. **Select the title changes.**

In the diff view, click the line numbers in the gutter to select the changes you want to keep. To select a whole code block, click the thin line just to the right of the line number. Select the code block next to line 5 by clicking the thin line next to line 5. After you select the code block, both lines labeled line 5 should be selected (selected lines show up as blue), as shown in Figure 7-6.

WARNING

You may be confused about why two lines are labeled 5 in the diff view. The numbers on the left represent what the file was originally named before you made the changes. The lines on the right represent the lines of the changed lines. Because I changed line 5, it's listed twice. Line 10 is a new line that didn't exist before, so it is listed only on the right.

4. **With those lines selected, enter a commit message and then click the Commit to Main button.**

As you can see in the bottom left portion of Figure 7-6, Desktop provides two fields for commit messages. Go ahead and enter **Change the title** into the summary and click the Commit to Main button.

Notice that the diff view updates to have the change only on line 10 (see Figure 7-7). That's because I committed the change on line 5.

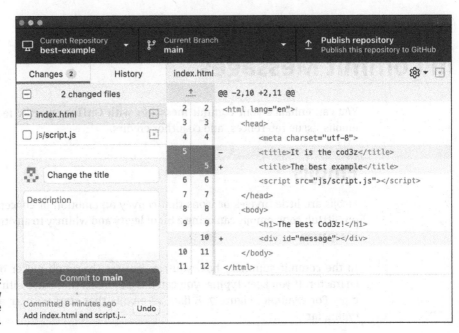

FIGURE 7-6: The diff view with one change selected.

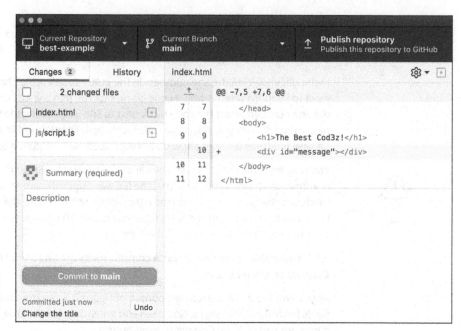

FIGURE 7-7:
The Changes tab after a partial commit.

If you're following the example in this chapter, make sure all the remaining changes are selected by clicking the check box next to the label 2 `changed files` until the check box is selected.

Using GitHub Conventions in Commit Messages

You can enhance your commit messages with GitHub-specific features, such as emojis, issue references, and coauthor credits.

Emojis

Emojis are little images or icons that convey an emotion or concept. Widely used on GitHub.com, emojis can bring a bit of levity and whimsy to an otherwise serious occupation.

In the commit summary box, you can initiate the emoji picker by typing the : character. If you keep typing, you can list all emojis that start with the letters you type. For example, Figure 7-8 lists all emojis that start with ar as the result of typing :**ar**.

FIGURE 7-8:
The emoji picker
listing emojis.

You can select the one you want with the arrow keys and then press Tab to complete it. Desktop then fills in the full text of the emoji, which in this case is :art:.

Issue references

GitHub also lets you reference an issue in a commit message with the format #123 where 123 is the issue number. Desktop has support for looking up an issue when writing a commit message. To try this out, create an issue ahead of time so that you can reference the issue in a commit message. As an example, I created an issue that describes the need to test the greeting created by a GitHub Action for new contributors to this repository. I reference that issue in this commit message.

To reference an issue in a commit message:

1. **In the commit description field, type** Fixes #.

   A few recent issues appear. If you don't see the issue you want to reference and you don't remember the issue number, you can start typing a word that's in the issue that you remember. For example, when I type **# greetings** an issue pops up (see Figure 7-9).

TIP

2. **Select the issue you want to reference and press Tab.**

   In this example, I selected issue 15. Desktop replaces #greetings with #15.

Giving credit to coauthors

Git doesn't support multiple authors directly. However, the Git community created a convention for specifying multiple coauthors in a commit that is now supported by GitHub.

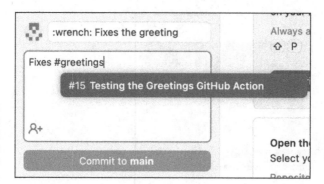

FIGURE 7-9:
A list of issues
with the word
"greetings" in
them, in this case
there is only one.

To give credit to coauthors:

1. **With the Desktop open, in the commit box with the Description label, click the little icon with a person and a plus sign in the bottom-left corner.**

   Desktop adds a textbox to enter a coauthor's GitHub username.

2. **Click the @ symbol to see a list of potential users, as shown in Figure 7-10.**

   GitHub lists only users who have access to the repository — for example, collaborators and org members (if the repository belongs to an organization).

   Just like the issue selector, you can also search by first name, last name, or username by appending a bit after the @. Press Tab, and Desktop replaces whatever you typed so far with the selected user's full username.

3. **To create the commit, click the Commit to Main button.**

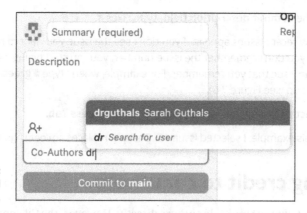

FIGURE 7-10:
A list of potential
coauthors.

To see your commit, click the History tab and click the commit you just created (see Figure 7-11).

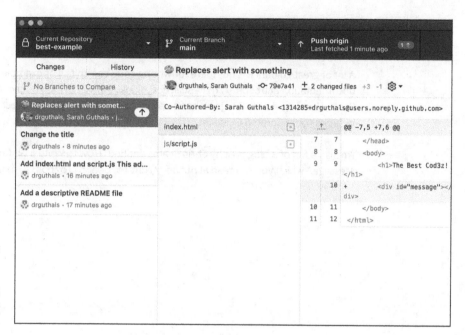

FIGURE 7-11:
The newly
created commit.

You can see in the commit message in the right pane that my username was replaced by the line

```
Co-Authored-By: Sarah Guthals <1314285+drguthals@users.noreply.github.com>
```

REMEMBER

That's the actual convention for specifying coauthors in Git commit messages. By using Desktop, you don't have to remember the exact format. You can just specify a username and let Desktop handle the rest.

Committing Code from Your Editor

Many editors have built-in support for committing code. Built-in support allows you commit code without having to switch to another application. The downside is that different editors have different levels of support for the various conventions you can use in a commit message.

But for quick and dirty commits, built-in support is very useful. Covering how every editor supports Git commits is out of the scope of this book, but you can see this in action with VS Code in Chapter 5. For other editors, refer to their specific documentation.

FOR MORE READING

A lot of great guidance is out there for writing good commits. For example, the Git documentation at https://git-scm.com/book/en/v2/Distributed-Git-Contributing-to-a-Project has a section on contributing to a project and includes some Commit Guidelines.

I'm also a fan of a blog post by Chris Beams entitled "How to Write a Git Commit Message," which you can read at https://chris.beams.io/posts/git-commit/.

Chapter 8

Working with Pull Requests

n Chapter 7, I note that a commit is the smallest unit of work with Git. On GitHub, the pull request is the main unit of work.

In this chapter, I explain exactly what a pull request is and how it pushes code to GitHub. I also describe the processes for opening, writing, and reviewing a pull request.

Understanding a Pull Request

The name pull request is confusing to some folks. "What exactly am I requesting to be pulled?" Good question. A *pull request* is a request to the maintainer of a repository to pull in some code.

When you write some code that you want to contribute to a repository, you create and submit a pull request. Your code contains some proposed changes to the target repository. A pull request is your way of offering these changes to the maintainer of the repository. It gives the repository maintainers an opportunity to review the changes and either accept them, reject them, or ask for more changes to be made.

Pushing Code to GitHub

To push code to GitHub, you need a repository. Open the repository of your choice. If you don't have a repository yet, Chapter 7 walks through creating a repository that you can use.

TIP

If you'd like to follow along with my example but you haven't completed the steps in Chapter 7, fork the repository `https://github.com/dra-sarah/best-example` and then clone your fork to your local machine. If that last instruction sounds like gobbledygook to you, you may want to review Chapter 6, which covers forking and cloning.

The first thing to do is create a new branch. Creating a new branch before writing new code is standard operating procedure with Git. I have a confession to make. I neglected to mention that in the example in Chapter 7. The example directs you to commit code directly to the `main` branch. That was a shortcut I suggested to keep things simple.

In this chapter, you do things the right way and work in a new branch. There's an important reason for this. A pull request doesn't consist of an arbitrary set of changes or commits. A pull request is always associated with a branch. In other words, a pull request is a request to merge one branch into another.

TIP

While it's true that a pull request can target any branch (except itself), the most common scenario is to target the main branch of a repository, typically named `main`.

This relationship between pull requests and branches is why you should create a new branch when starting new work. I name the branch `new-work` for this example, but feel free to name it whatever you want by replacing `new-work` with your own branch name in the following command:

```
$ git checkout -b new-work
```

Now that you have a branch, you need to create a commit in that branch. For this example, the specific contents of the commit are not important. You can choose any file and make some edits to the file, such as adding some text to the end. Or if you're following along with the repository I created in Chapter 7, manually edit the `README.md` file or run the following command to append some text to the end of the file:

```
$ echo "\n## Installation" >> README.md
```

Now that you have a file with some changes, commit those changes. You can use the following command, for example, to commit all your changes. The commit message is not important here. The important thing is to have a commit in a branch to work with that is not in the main branch.

```
$ git add -A
$ git commit -m "Add text to the README"
```

Now push this new branch to GitHub.com (replace new-work with your branch name):

```
$ git push -u origin new-work
```

The git push command tells Git to push local commits to a remote repository. The -u flag specifies where to push it — in this case, to a branch also named new-work on the remote named origin.

The -u flag is only needed the first time you push a new branch to the server. From that point on, the new-work branch on your local machine is associated with the new-work branch on GitHub.com and any subsequent pushes to that branch do not need the flag.

Opening a Pull Request

Before you can open a pull request, your GitHub.com repository must have at least one branch other than the default branch. If you follow the steps in the earlier section "Pushing Code to GitHub," you have a branch that is not yet merged into main. In my case, the branch is named new-work branch, but you may have named yours something else. Visit the repository on GitHub.com to open a pull request.

GIT ALIAS FOR OPENING GITHUB.COM FROM THE TERMINAL

Often when working with a repository in the terminal, you need to jump to the repository on GitHub.com. I have an alias just for this purpose. Chapter 6 covers aliases in more details and how to add them. You can run the following in the terminal to add a new git browse alias.

(continued)

(continued)

```
$ git config --global alias.browse '!open `git config remote.origin.url`'
```

To use the alias, just run the following in the terminal from within your repository directory.

```
$ git browse
```

This code launches your default browser and navigates to your repository's origin URL. This alias assumes you're using https for your git remotes and not SSH.

When you visit the repository on GitHub.com, you see a new message in the at the top of the Code tab, as shown in Figure 8-1.

FIGURE 8-1:
Repository home page with recently pushed branches listed.

Click the Compare & Pull Request button to navigate to the Open a Pull Request page, as shown in Figure 8-2. The target branch (the branch you want to pull your changes into) is the default branch for the repo. Your branch is listed next to the target branch, and a status of whether your branch can be merged into the target branch is next to that. The pull request title is the same as the most recent commit message — in this example Add text to README — and your description is blank. Figure 8-2 shows an example description.

Click the Create Pull Request button, and you go to the repository with the new pull request open. Here you can see that the account dra-sarah, or any accounts who have been added as collaborators to this repository, can assign reviewers, assign assignees, apply labels, connect to a project, or add to a milestone. Figure 8-3 shows the pull request opened on the dra-sarah repository.

Status   **Target repository and branch**   **Your repository and branch**

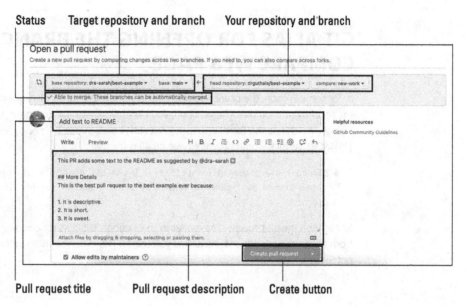

FIGURE 8-2:
The Open a Pull
Request page.

Pull request title   **Pull request description**   **Create button**

Reviewers and option to convert to draft

Assignees

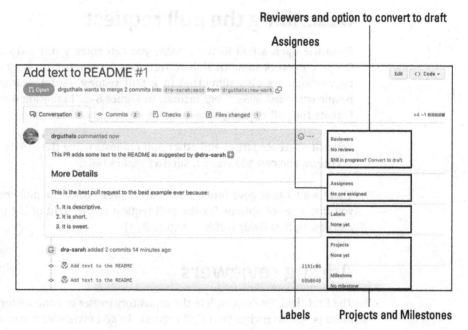

FIGURE 8-3:
The pull request
opened on the
dra-sarah/
best-example
repository.

Labels   **Projects and Milestones**

TIP

You can change the default branch by choosing the Settings ⇨ Branches section of
your repository.

Describing the pull request

From the Open a Pull Request page, you can enter a summary and description. Chapter 7 covers some GitHub conventions for commit messages. Most of those conventions are also supported in a pull request — for example, mentioning people using the @USERNAME format. In Figure 8-2, I mention @dra-sarah. When I create the pull request, @dra-sarah receives a notification.

You can reference issues and other pull requests using the #ISSUEID format. And, of course, you can add emojis, such as :sparkles:.

There's a lot that goes into a good pull request. Once the pull request is opened, you find a set of options for the pull request to the right of the pull request title and description fields (refer to Figure 8-2).

Adding reviewers

The first field, Reviewers, lets the repository owner or contributors specify one or more people to review your pull request. To add reviewers if you are the owner or a contributor on a repository:

1. **Click Reviewers to see a list of people you can mention.**

   For repositories with a large number of users, you can start typing to filter the set of users.

2. **Click each user to add them to the list of reviewers.**

When you add a reviewer, they're immediately notified when you finish creating the pull request.

TIP

If you aren't ready to have reviewers start looking at your changes, you can convert your pull request to a draft. This allows you to continue pushing changes, easily view the diff, but not get comments or critiques of your work in progress. You can learn more on the GitHub pull request docs at `https://docs.github.com/pull-requests`.

Specifying assignees

The next option after Reviewers is an option to specify assignees. An *assignee* is the person who should take action on the pull request. Often, a pull request represents a work in progress and not the final result of some work. If more work needs to be done on a pull request, you'd assign the pull request to the person that should do the work.

To specify assignees:

1. **Click Assignees to see a list of assignees.**

The assignee dialog box works just like the reviewers dialog box, described in the preceding section. It allows you to select one or more assignees.

2. **Click each user to add them to the list of assignees.**

TIP

In most cases, it's best to just assign one person who will be responsible for the next step. Assigning one person reduces the chances that multiple assignees all think the other assignees are responsible for the work.

Specifying labels

Labels provide convenient grouping and context to help you decide what to work on next or what to review next.

The set of labels you can use on issues and pull requests are the same, but some labels make more sense for issues than pull requests and vice versa. For example, many repositories have a "ready for review" label specifically for pull requests.

Specifying projects and milestones

The last two options allow you to specify the project board and milestone that this pull request belongs to. Chapter 3 covers projects, and Chapter 11 covers milestones.

Writing a Great Pull Request

Writing a great pull request is a bit of an art. For an open source project, much of the project's communication with people occurs within pull requests.

If you're contributing to a project, your pull request is your chance to make a strong case for why your code should be pulled into the main branch. Make sure to put your best foot forward.

Knowing your audience

Before you write a single word, understanding your audience is helpful so that you can focus your words on what is most useful. A pull request may serve many audiences. Keeping all your audiences in mind is important, but your primary focus is on the folks who will review and make a decision on whether your pull request will be merged. You want to make their lives easier as they tend to be very busy.

Even though the project maintainers are your primary audience, you should never forget that many others may read the pull request. For an open source project, that audience may be the entire world. So keep your language respectful, friendly, and inclusive.

WARNING

It's pretty common to have someone write a pull request in a fit of anger and later regret the words they use. So if you happen to be rage coding, take a moment to cool down and gather your thoughts before creating the pull request.

Making the purpose clear

Make sure to be concise and informative. For example, the summary should make the purpose of the pull request clear. The summary is the only part shown on the page that lists pull requests. It needs to be scannable.

Here are some examples of good pull request summaries:

>> Adds the About page to the website

>> Minimizes boilerplate setup code for JavaScript libraries

>> Extracts and isolates error handling from GitStore internals

Here are some bad examples taken from my own repositories:

>> Teams are forever

>> Typo

>> Small changes

TIP

The description should provide a bit more explanation about the purpose of the pull request. Don't write a book, but do make it clear what the pull request attempts to accomplish.

Keeping it focused

Much like a commit, a pull request should not contain a bunch of unrelated changes. A pull request may consist of multiple commits, but they should all be related to the task at hand.

You can often tell that a pull request is doing too much when writing a concise description of what the pull request accomplishes is difficult.

Even if the pull request is focused on a single major change, keep the pull request to a manageable size. Reviewing a very large pull request is difficult.

If the pull request addresses a very large task, break down the task into smaller steps and submit pull requests for each step.

Explaining the why

The previous section, "Keeping it focused," focuses on what the pull request does. You also need to explain why you're taking on this work. The pull request description is an opportunity to provide links to other documents that provide more context. You can't assume everyone will be familiar with the history.

If you have a lot of context to share, you can provide a brief summary followed by more details within a `<details>` tag. For example, if you add a pull request comment with the following text:

```
The reason we're embarking on this work is due to compliance reasons.
<details>
## More Details
I don't want to bore everyone with all the nitty gritty details, but for those
    who are interested, keep on reading...
</details>
```

GitHub displays the details section collapsed by default, as shown in Figure 8-4.

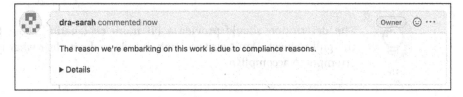

FIGURE 8-4:
Details section
collapsed.

Click the details label to expand the details section, as shown in Figure 8-5.

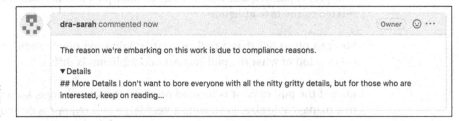

FIGURE 8-5:
Details section
expanded.

A picture is worth a thousand words

GitHub supports adding images to a pull request description by dragging and dropping an image. When you drag and then drop the image on the text field, GitHub uploads the image and replaces it with the Markdown for rendering an image. Figure 8-6 shows an upload in progress after I dragged an image into a pull request comment field.

FIGURE 8-6:
Uploading an
image to a
pull request.

Visit `https://github.com/dra-sarah/best-example/pull/1-issuecomment-1244931357` to see this comment. It's very meta as it's a screenshot of the same repository it's a comment on.

TIP

If an image is worth a thousand words, an animated gif is worth even more. If you can create an animated gif that demonstrates the behavior change introduced by the pull request, adding that gif to the pull request is usually very appreciated by those who review it.

Including a call to action

You need to be very clear about what feedback you want from others on the pull request. For example, if the pull request is a work in progress, make that clear from the start so that people don't waste time reviewing a pull request that isn't ready for review.

To make that clear, follow the conventions of the repository. You can find out the conventions by orienting yourself with the repository as described in Chapter 5.

Following the conventions is important so that others know what is expected of them with respect to your pull request.

WRITING A PERFECT PULL REQUEST

The previous suggestions describe writing a great pull request. It's not comprehensive. For more suggestions, check out a blog post on the GitHub blog entitled "How to write the perfect pull request" at `https://blog.github.com/2015-01-21-how-to-write-the-perfect-pull-request/`.

Reviewing a Pull Request

Pull requests have two parties involved: the person who writes and opens the pull request, and the person (or people) who reviews it. In this section, you put yourself in the position of a reviewer of a pull request.

REMEMBER

Reviewing a pull request is a very active activity. It takes a lot of focus and attention to do it well. It doesn't serve the folks submitting pull requests if you just take a cursory look at it, comment LGTM (Looks Good To Me), and don't provide quality feedback.

Sometimes, a cursory look is all you have time for. In that case, make it clear what you did and didn't review and suggest that someone else provide a more detailed review.

When someone adds you as a reviewer, you'll receive a notification about the request, typically via email or on your notification bell at the top right of GitHub.com. If you visit the pull request, you see a banner message, as shown in Figure 8-7.

FIGURE 8-7:
Message requesting your review on this pull request.

WARNING

Don't click the Add Your Review button just yet. That takes you to a dialog box to write a write-up of your review. Only do that when you're ready to complete your review.

Reviewing the Conversation tab

When you review a pull request, the first thing to do is read through the contents of the Conversation tab for the pull request. Make sure that you understand the purpose of the pull request and why it's necessary.

At the bottom of the conversation is a section that displays the checks that GitHub runs against the repository if there are any. If no checks are set up, you see the message `Continuous integration has not been set up.` The status of these checks is the next thing you should check. Many repositories have several checks, such as does the code compile, did all the tests pass, and so on.

If any of these tests fail, you shouldn't spend any more time reviewing the code. The person submitting the pull request can see that their pull request doesn't pass all the checks, and it's up to them to fix it.

TIP

Even though the person submitting a pull request should be able to see that the checks have failed, sometimes they don't stick around long enough after creating the pull request to see that. You may want to add a gentle note that mentions the person submitting the pull request and informs them that the checks have failed and they should try fixing the problems and push their changes to the pull request branch again.

Figure 8-8 shows an example from the GitHub Desktop open source project of a pull request that fails the continuous integration (CI) builds.

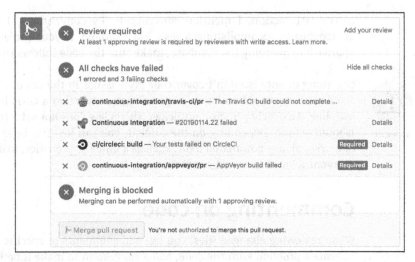

FIGURE 8-8:
Pull request with
failed checks.

Reviewing the changed files

Assuming you're happy with the contents of the Conversation tab and all the checks pass, it's time to get into nitty gritty and review all the file changes. Click the Files Changed tab to see all the changes in the pull request.

At this point, you should review the code for things like

>> **Clarity:** Is the code easy to read and understand? Could it be made more clear? Are there appropriate comments throughout the code? Obtuse, hard-to-understand code becomes a maintenance nightmare down the road.

>> **Correctness:** Does the code do what the pull request says it does? Are there any glaring bugs? Are there any errors of omission? Are there any tests missing?

>> **Security:** Related to the previous item, a security review requires a specific mindset. Ideally, you work with security experts who can help review the code for security. The idea here is to think about all the ways a bad actor could attempt to attack the code. There are many frameworks for doing security review, such as STRIDE. You should also think about how bad actors can use the code to harm other users. Does the code protect users privacy? Does it ask for consent to take actions on behalf of users?

>> **Conventions and idioms:** Just because code is correct, it doesn't mean it's necessarily idiomatic. A code review is a good place to teach and learn conventions and idioms specific to a project.

In the last section, I mention conventions. By conventions, I mean common approaches to accomplishing a task. For example, if your project has a certain approach to querying the database, make sure the code follows that approach.

WARNING

One thing to note is I don't cover code style issues in the list of things to review. A code review should not cover nitpicks such as whether a curly brace goes on its own line. An overly pedantic nitpicky code review does not set a friendly and collaborative tone. Depending on the context, you can just fix these things yourself or better yet, use automated tools, such as a linter or prettifier, to do it. You'll save everyone a lot of time and headaches.

Commenting on code

When reviewing changed files, you can add comments to specific lines of code to indicate a problem with the code, add a suggestion to make it better, or celebrate someone's awesome code writing skills with a :sparkle:.

TIP

Positive and encouraging comments set for a welcoming and collaborative tone. Maintainers often forget how daunting it can be to contribute code to a project for the first time. Don't be stingy with the :sparkle: emojis!

To comment on code:

1. **Hover your mouse over the line of code.**

   A blue square with a plus sign appears next to the line number.

2. **Click the square to reveal a comment form for that specific line of code, as shown in Figure 8-9.**

   The comment form supports the same things issues and pull requests do, such as mentions, Markdown, and, of course, emojis.

   At this point, you can choose to click Add Single Comment or Start a Review.

   Clicking Add Single Comment immediately adds your comment to the pull request and sends any notifications. This can be useful when making a quick one-off comment. In most cases, I recommend against this approach as it doesn't lead to well considered and thoughtful code reviews.

3. **Click Start a Review instead to create a review and add the comment as a pending comment to the review, as shown in Figure 8-10.**

   The pending label indicates that you're only one who can see the comment so far. By starting a review in this manner, you can continue to add (and edit) pending comments as you review the code and only publish them when you're completely satisfied with the review.

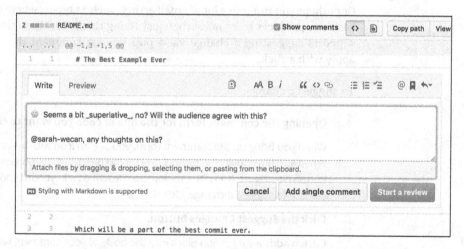

FIGURE 8-9:
Pull Request
Comment form
for a line of code.

TIP

The reason I recommend starting a review as opposed to adding single comments as you go leads to a more coherent and helpful code review. A common occurrence when reviewing code is that something you read later in the review makes you realize a comment you made earlier should be updated or even deleted.

A review lets you make those adjustments before you send anything to the author. This option gives you an opportunity to review your review before publishing it.

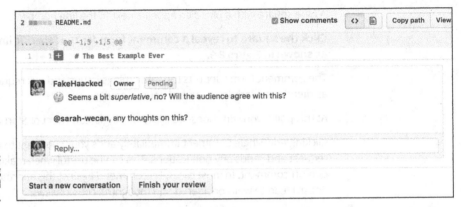

Suggesting changes

Sometimes as you review code, you come across a section of code where you feel like it'd be faster to just fix the code than try to explain in words what should be fixed.

Or perhaps you run into a lot of small errors, such as typos, where commenting on each one produces more noise than just fixing them. For these situations, GitHub supports suggesting a change via a pull request comment that the author can apply with a click.

To suggest a change:

1. **Opening the comment form for the line of code you want to change.**

When you bring up the comment form, you see an icon with a + and - symbol. When you hover over the symbol, the tooltip describes the purpose of this button (see Figure 8-11). The tooltip also describes a keyboard shortcut you can use to suggest a change, CMD–G.

2. **Click the Suggest Changes button.**

GitHub adds a suggestion block into the body of your comment with the current contents of the line you want to change.

3. **Change the contents of the block to suggest what the final result should be.**

Figure 8-12 shows an example where I suggest a change to add the word "Instructions" to the line.

4. **Click the Add Review Comment button to save your comment to this review.**

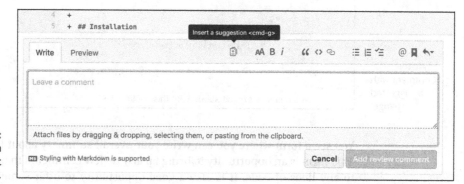

FIGURE 8-11:
Comment form
with the Suggest
changes button.

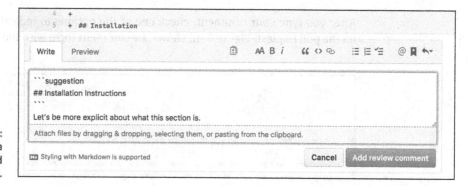

FIGURE 8-12:
Comment with a
suggested
change.

TIP

Anything you write outside of the suggestion block in the comment is not included in the suggested change. This lets you provide a reason for the suggested change. Except for truly simple or self-explanatory suggestions, I recommend always providing this extra context.

After you create a comment with a suggested change, GitHub displays the change as a small diff within the comment. If the person viewing the suggestion has commit access, they see an option to commit the suggested change (see Figure 8-13). They will also thank you for saving them time!

Finishing the review

After you have made all your suggestions, you can finish the review so the author receives all your valuable feedback and can start to address it. To finalize the review, click the Finish Your Review button at the bottom of any pending comment (refer to Figure 8-10).

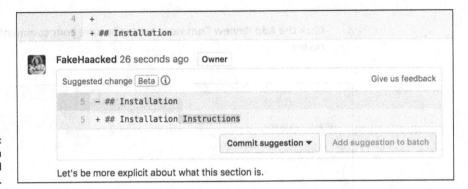

FIGURE 8-13:
Comment with
a suggested
change.

You see a form where you can write your overall summary about the pull request. This form is an opportunity to bring up any review comments that are not specific to any lines of code. It is also a good opportunity to offer some general praise, raise broad concerns, suggest follow-up actions, and so on.

After you type your comment, check one of the options to indicate your position on the pull request. Figure 8-14 shows the comment form with the review options.

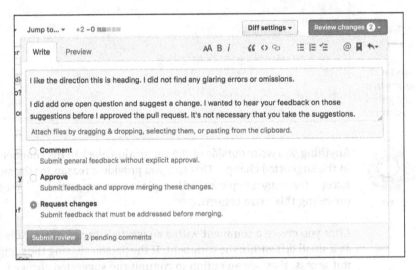

FIGURE 8-14:
Comment with
a suggested
change.

TIP

If you're reviewing your own pull request, the only option available is to Comment on the pull request.

Your choice may determine whether this pull request can be merged into the default branch. Whether your choice blocks a merge depends on how the branch protection rules in place for the repository. For example, some repositories may require a certain number of approved reviews before the pull request can be merged. Choose Settings⇨ Branches to manage branch protection rules.

Reading More about Pull Requests

Pull requests and code reviews are a very important part of the software development lifecycle. As such, a lot of great advice is out there for doing these things well — more advice than I can offer in a single chapter. Here are two articles to read to help you get the most out of pull requests and code review.

» **Building an Inclusive Code Review Culture:** How you review code is a reflection of your overall culture. This article describes an approach that is inclusive and collaborative.

```
https://blog.plaid.com/building-an-inclusive-code-review-culture/
```

» **Code Review Like You Mean It:** This article is a brief write-up about the efficacy of code review along with some tips on how to do them well.

```
https://haacked.com/archive/2013/10/28/code-review-like-you-mean-it.aspx/
```

4

Managing and Contributing to Large Projects

Chapter **9**

Exploring and Contributing to OSS

or developers, GitHub is the greatest source of treasure ever created, if you know where to look. An open source repository exists for every possible need out there. This cornucopia of choice can be overwhelming at first.

Going beyond just making use of open source software (OSS), contributing to open source is a fantastic way to continue your development as a software developer. It gives you the opportunity to work with technologies that you may not otherwise work with in your day job. It connects you with a large community of developers working on a diverse array of challenges. Many of these developers are happy to share their knowledge with folks looking to learn something new.

In this chapter, I look at ways to explore the range of open source software GitHub has to offer. I not only look at ways of discovering repositories you may want to use in your own projects, but also provide tips for finding repositories you may want to contribute to.

Exploring GitHub

The Explore page on GitHub, located at `https://github.com/explore`, is a great starting point to discover repositories, topics, collections, and OSS projects that may align with your interests. It contains a mix of human and algorithmically curated content. The Explore page displays a selection of repositories based on your interests. This selection is an algorithmically curated page of recommendations specific to you. Figure 9-1 shows a portion of the Explore page.

GitHub employs machine learning techniques to generate a list of recommendations based on repositories you've starred, contributed to, and viewed. The recommendations also factor in the people you follow on GitHub.

TIP

To improve the recommendations, be mindful of the repositories you star and the people you follow.

In the next several sections, I describe each section of the Explore page and its significance. Where it makes sense, I cover what you need to do to make your repository a candidate for inclusion in the section.

Exploring topics

The Popular Topics section lists the most popular topics on GitHub. A *topic* is a user-applied category for a repository. A topic gives people more information about what a repository is about. A repository owner can specify multiple topics for a repository. The concept is very similar to tags or issue labels.

Figure 9-2 shows the list of topics for the `thewecanzone/GitHubForDummies Readers` repository. Admins for the repository see a Manage Topics icon that lets them add or remove topics for the repository.

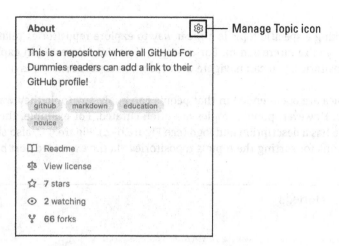

Manage Topic icon

FIGURE 9-2:
List of topics for
a repository.

When you add topics, you type a portion of a topic name, and GitHub offers suggestions based on the topics that others use within GitHub. Figure 9-3 shows an example where I typed novice, and GitHub suggests other topics with the word novice in them.

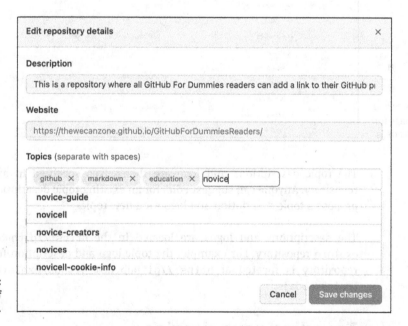

FIGURE 9-3:
A list of
suggested topics.

Visit the Topics page at `https://github.com/topics` to see the most used topics on GitHub.

Clicking a specific topic is a great way to explore repositories related to a subject that you're interested in. For example, if you're interested in exploring Node.js repositories, you can navigate to https://github.com/topics/nodejs.

Topics are open-ended in that people can apply any topic they want to a repository. However, popular topics are often curated. For example, the Node.js topic page has a description and logo (see Figure 9-4). Figure 9-4 also shows the many options for sorting the topic's repositories via the expanded Sort button.

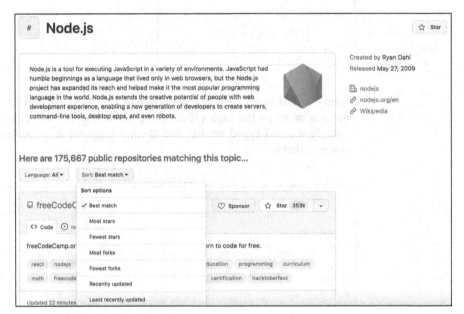

FIGURE 9-4: Topic page for Node.js.

The topic descriptions and logo are themselves specified in an open source repository. Anyone can suggest edits for an existing topic description. Anyone can propose a topic description and logo for a new topic.

The descriptions and logos are located in the https://github.com/github/explore repository. For example, the topic logo and description for the node.js repository is located at https://github.com/github/explore/tree/main/topics/nodejs.

Trending repositories

The next section on the Explore page includes a list of the top 25 trending repositories. Click the heading to visit https://github.com/trending, a page where you can discover repositories and people that are trending across GitHub.

REMEMBER

Why only 25 trending repositories? To have more than 25 trending repositories would dilute what it means to be trending. Also, it takes a lot to compute trending repositories, so limiting it keeps the cost low.

This page lets you filter trending repositories based on the primary language of the repository. If you're interested in the trending JavaScript libraries, click the Other Languages button and select JavaScript.

The Trending button at the top of the page defaults to today, but you can click it to see what's been trending for the week and month.

To determine what's trending, GitHub looks at a variety of data points, such as stars, forks, commits, follows, and pageviews. GitHub weighs these data points appropriately and factors in how recent the events were, not just total numbers.

Exploring collections

The next three sections are curated collections. At the time of writing, they are currently Pixel Art Tools, Game Engineers, and Made in Brazil.

Each collection is fully curated by a human. A collection may contain a list of web pages and repositories. The goal is to be a great starting point for learning a particular subject in depth by listing websites for further reading and repositories with related code.

For example, to edit The Learn to Code collection, suggest an edit to this file in the `github/explore` repository:

```
https://github.com/github/explore/blob/main/collections/learn-
    to-code/index.md
```

Exploring events

The Events section lists a selection of upcoming GitHub affiliated events, such as GitHub Universe (GitHub's yearly flagship conference), GitHub Satellite (smaller GitHub community events hosted around the world), and All Things Open (an event promoting open source).

Check this page often to find out about future events, which can be a great way to connect with the larger GitHub community.

Exploring GitHub Sponsors

The GitHub Sponsors section takes you to `https://github.com/sponsors/explore` where you can find a list of all of the open source projects and maintainers that create dependencies you have in your repositories. This is a great way to monetarily give back to the projects and people that make what you do possible.

Getting by with help from your friends

Exploring what others on GitHub are up to is a great way to discover interesting new repositories. Whether it's your friends or other people you admire, be sure to pay attention to what they're up to on GitHub. One way to do so is through stars.

TIP

As you explore repositories on GitHub, be sure to star the ones that pique your interest. To see your starred repositories, go to `https://github.com/stars`. Starring a repository not only is a good way to bookmark a repository for later exploration, but it also is a nice way to show a repository some recognition.

On the right of the stars page is a grid of avatars for your friends on GitHub. Click a friend to see the repositories they've starred. Starred repositories often surfaces interesting repositories you may not have otherwise noticed.

You can also visit the profile page for your friends or other developers who you admire and look at their pinned repositories. *Pinned repositories* are repositories that a person explicitly chooses to feature. Figure 9-5 shows the pinned repositories for a GitHub user.

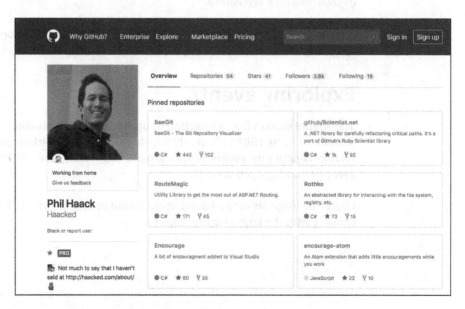

FIGURE 9-5:
A profile page.

Finding Places to Contribute

Some of you may be ready to move beyond just using open source code and are ready to hone your skills by contributing back to open source. How do you find a project to contribute to?

Exploring interesting repositories on GitHub comes into play when looking for a place to contribute. The only difference is that you'll have to narrow your exploration down even more when looking for a place to contribute.

The first thing to ask yourself is what type of repository do you want to contribute to. Answering this question can help narrow and guide your search.

There are many valid reasons for contributing to open source. One or more of the following may apply:

>> I want to get involved in OSS in general and learn how to contribute.

>> I want to help improve a project that I use in my day-to-day work.

>> I want to support a project that does good in the world.

>> I want to expand my skills by working in a technology I don't normally get to.

>> I'm just bored and want to contribute to something cool.

No matter your reason for contributing to open source, a good place to start when approaching a new repository is to look for low-hanging fruit. Many repositories have some manner to indicate issues that would be great for first-time contributors. Often, they apply a label, such as `help-wanted` or `up-for-grabs`.

In fact, some websites scour issues labeled as such and make them searchable to those looking to get started with open source. The Up For Grabs website at `https://up-for-grabs.net` is one example.

The site has a search tool for filtering issues by project, label, and tag. Figure 9-6 shows an example where I'm looking at all projects that are tagged with `javascript`.

Several repository results are displayed along with the label that matched the filter. Click the label for the repository result to see the specific issues for that repository with that label. Filtering issues is a good way to find a concrete issue to work on as a beginner.

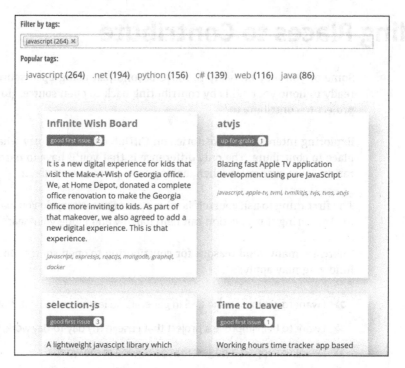

A GitHub topic named `help-wanted`, which is at page `https://github.com/topics/help-wanted`, lists repositories that are looking for help. When you visit a project that needs help, visit the Issues page to look for potential issues you can help with.

For example, Figure 9-7 shows the issue tracker for the Visual Studio Code open source project. Figure 9-7 shows the issues filtered by the `good first issue` label.

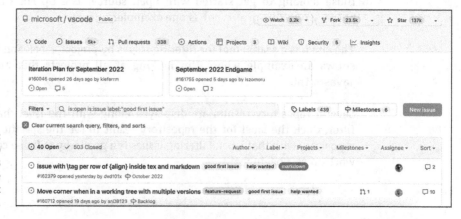

Surveying a Project for Contribution

Suppose that you decided that you want to contribute to a particular project. What are the next steps? For this example, I look at the Visual Studio Code repository at `https://github.com/microsoft/vscode`.

Reading the contributing guide

Before you embark on contributing to an open source project, you should always read through the contributing guidelines first.

TIP

The contributing guidelines message is controlled by a convention. Add a file named `CONTRIBUTING.md` to the root of the repository, a folder named `docs`, or a folder named `.github` to specify contributing guidelines for a repository.

Visual Studio Code's contributing guidelines are located at `https://github.com/Microsoft/vscode/blob/main/CONTRIBUTING.md`.

Like many contributing guides, the guide provides a high-level overview of how to make contributions to the project. It notes that contribution is more than just writing code. To emphasize that philosophy, the guide starts by letting you know where to ask questions about the project. The guide addresses the key topics:

>> Where to ask questions

>> Where to provide feedback

>> Where and how to report issues

>> Details about their automated issue management

>> A link to a guide on how to contribute code

This format is pretty common for contributing guides. Sometimes, a smaller project also includes how to contribute code in the main document, but for larger or more complex projects, contributing code is a big topic.

Reading the contributing code guide

The VS Code project keeps its code contribution guide in a wiki located at `https://github.com/Microsoft/vscode/wiki/How-to-Contribute`.

REMEMBER

If you plan to contribute code, read through this guide carefully.

The guide has the key information you need to code on the project and is structured in a fashion pretty common among open source projects:

>> Building the code

>> Running the code

>> Debugging the code

>> Running the automated tests

>> Running automated code analysis, such as linting

The next section covers the code contribution workflow:

>> Following the branching strategy

>> Creating pull requests

>> Packaging code for distribution

Reading the code of conduct

The contributing guide focuses on the mechanics of making a contribution. Many projects now also include a CODE_OF_CONDUCT.md file. This file lays out expectations for those who participate in the project. It sets the behavioral norms for the project and a resolution process in cases where violations of the norm occur.

VS Code uses the Microsoft Open Source Code of Conduct located at https://opensource.microsoft.com/codeofconduct.

TIP

Adding a code of conduct to your own repository is easy. If you add a new file on GitHub using the browser and name the file CODE_OF_CONDUCT.md, GitHub displays a code of conduct selection drop-down list with two choices: the Contributor Covenant and the Citizen Code of Conduct.

You can also visit the community page for any public repository to add a code of conduct. Go to the community page by appending /community to the end of the repository URL. For example, type https://github.com/microsoft/vscode/community to see whether the project has a code of conduct. If it doesn't, the page gives you the opportunity to propose one.

Setting Contributor Expectations

After you've read the code of conduct and contribution guides, you're ready to dive in and make some contributions. Now's a good time to make sure that you have clear expectations for how the process will work.

REMEMBER

These general guidelines for contributors may not be spelled out in a contributing guide, but are more the result of collective wisdom gathered from working in open source for a long time.

They won't fix every issue

Many, if not most, projects have limited resources, and their priorities may not align with your priorities. It's important to keep all that in mind when you file an issue. On the one hand, opening a good issue takes time and energy. A well-written issue that thoroughly describes a problem with clear steps to reproduce the issue is very valuable to a project, so it's understandable that people who write issues feel invested in them.

REMEMBER

On the other hand, the fact that an issue may not be fixed is why I say that writing an issue is a contribution. It's not an exchange of one thing for another, but it's a gift. And as such, you can't expect much in return. Good project maintainers thank you for filing an issue and note whether they know of any workarounds at minimum, but they do not owe you a fix. Do not take it personally if an issue you file is labeled as won't fix.

They won't merge every pull request

Submitting a pull request is the culmination of a lot of work. To submit a proper pull request, a developer had to spend the time to get the code working on their own machine, understand the code well enough to make a change, write the change, and then submit it.

Being disappointed when the repository maintainers then reject the pull request is understandable. Remember, though, they're under no obligation to accept your pull request. Yes, you put a lot of time into the pull request, but they'll have to own the code change for the lifetime of the project.

REMEMBER

Having a pull request rejected should be a very rare occurrence. You and the project maintainers can do a lot beforehand to avoid a rejected pull request. The first step starts with communication. Before you start work to resolve an issue, comment on the issue with your intentions. Indicate your general plan of attack.

Make sure that someone related to the project responds and agrees with your general approach. This communication reduces the likelihood that you go off and work on something in a matter completely contrary to the project.

Don't stop your communication there, though. Keep it going. As soon as you have a commit or two, push it to a pull request and make sure the pull request is clearly marked as a work in progress. Marking the pull request as a work in progress lets you continue to get feedback as you go and ensure that you're on the right path. It minimizes the time spent going down the wrong path.

Finally, make sure that you're following all their contribution guidelines, style guidelines, and so on. This is their project, and they will own the code if they choose to merge your pull request in. This is not the time and place to try to advocate for your way of working or your personal code style.

They don't owe you anything

One of the biggest challenges of being a maintainer of a popular open source project is the sense of entitlement expressed by many people who use the software.

REMEMBER

In most cases, the maintainers are all volunteers working on the project on the side. This isn't always the case. Sometimes, they're paid employees working on an open source project. But chances are, you are not the one paying them. They don't owe you a feature. They don't owe you a bug fix. They don't owe you anything. And you should treat them accordingly.

Of course, the maintainers of a well-run project try to go out of their way to accommodate people who participate in a project. It's only natural that they want most people involved to be happy. Open source projects benefit from mind share and more contributors, so it's often in their best interests to not take a hard line and try to get you that feature if it fits their roadmap and priorities and isn't too much trouble. So it's not wrong to ask for things. It's not wrong to make a strong case for things you want. But at the end of the day, you don't pay them, and they don't owe you a thing.

Keeping Tabs on a Project

When you're a contributor to a project, it's good to keep tabs on how the project is doing. One way is through GitHub notifications. Chapter 1 covers how to manage notification settings. GitHub can send notifications for new issues, new comments on pull requests, and so on. It's important to manage your notification settings based on your interest level in the project.

While GitHub may be the hub of open source, a lot goes on outside of GitHub related to an open source project. You may want to look at the places where a project lives outside of GitHub. For example, many projects have a Twitter account you can follow. Some projects have a public Slack instance where you can chat with the maintainers and others who use the project. You also may want to read the blogs of the maintainers of those projects to keep on top of the latest developments. Many projects and project maintainers still publish RSS feeds, an ancient technology that makes it possible to keep up with a site.

Chapter **10**

Starting Your Own OSS

There comes a point in many developers lives when they have their own code to share with the world. Many different approaches and motivations may drive a person to share code. In some cases, a person may write some code they think others will find interesting and just "toss it over the wall" without any intent to take contributions back. On the other end is the case where someone wants to start a real movement with large numbers of contributors.

Whatever your motivation, the only requirement for open source software (OSS) is to choose a license that complies with the Open Source Definition (OSD). You can find the definition at `https://opensource.org/osd`. The Open Source Initiative (OSI) has a process for approving licenses, and its site lists a plethora of licenses.

But for the life of most open source software, choosing a license is just a starting point. A lot more goes with managing an open source project. In this chapter, I cover what it means to start an open source project on GitHub, maintain it, and if it comes to it, sunset it.

Creating an Open Source Repository

When you create a repository with the intent to make it open source, it's common to make the repository public and select a license during the repository creation process.

Sometimes you want to defer those choices to later. Perhaps you want to set up your repository right before you make it public to the world. Or you may want to spend more time thinking about the license. The following sections focus on the scenario of turning an existing private repository into an open source one. I assume that you've already created a repository with a README.md file, but didn't make the repository public nor did you choose a license. (If you haven't, see Chapter 3.)

Adding a license

The only requirement that software has to be considered open source is to have an open source license. GitHub makes adding a license to an existing project easy.

Figure 10-1 shows the home page for a private repository without a license. To add a license:

1. **Click the Create New File button.**

    After you click the button, GitHub prompts you to name the file.

2. **Type your filename.**

    For this example, I named mine LICENSE. GitHub notices you're attempting to create a license file and helpfully adds a button to the right labeled Choose a License Template, as shown in Figure 10-2.

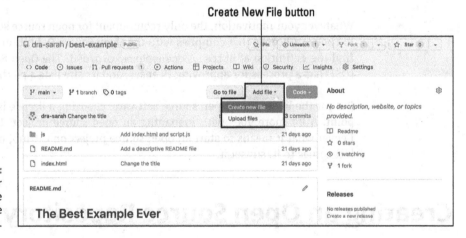

Create New File button

FIGURE 10-1: Home page for my soon-to-be open source repository.

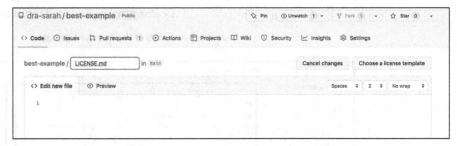

FIGURE 10-2:
Choose a
License Template
button magically
appears.

3. **Click the Choose a License Template button to see a list of license templates.**

   The three most common licenses used on GitHub are listed first in bold, as shown in Figure 10-3. Chapter 3 provides a brief guide on choosing a license, but this page has a link to a more detailed guide to help you pick which license best fits your project.

4. **After you know which license you want, click the license to view information about the license.**

   This page includes a high-level summary with bullet points about the key traits of the license. This overview helps you gain an understanding of the license without having to wade through all the legalese.

   Of course, the page also presents the full text of the license for those who relish legalese. On the right are fields required by the license.

5. **Enter your information in the fields to customize the license to your situation.**

   You can see an example of this page in Figure 10-4.

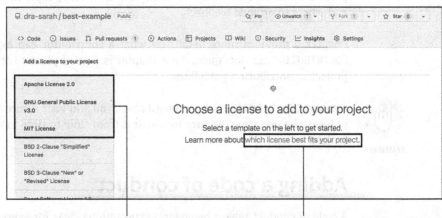

FIGURE 10-3:
The license
chooser page.

The three most popular licenses on GitHub   Link to more information on choosing a license

License template fields and button to move to next step

High-level summary of the license attributes

Add a license to your project

Apache License 2.0

GNU General Public License v3.0

MIT License

BSD 2-Clause "Simplified" License

BSD 3-Clause "New" or "Revised" License

Boost Software License 1.0

Creative Commons Zero v1.0 Universal

A short and simple permissive license with conditions only requiring preservation of copyright and license notices. Licensed works, modifications, and larger works may be distributed under different terms and without source code.

Permissions	Limitations	Conditions
✓ Commercial use	✕ Liability	ⓘ License and copyright notice
✓ Modification	✕ Warranty	
✓ Distribution		
✓ Private use		

This is not legal advice. Learn more about repository licenses.

To adopt **MIT License**, enter your details. You'll have a chance to review before committing a *LICENSE.md* file to a new branch or the root of your project.

Year ⓘ
2022

Full name ⓘ
dra-sarah

Review and submit

MIT License

Copyright (c) Year Full name

Permission is hereby granted, free of charge, to any person obtaining a copy of this software and associated documentation files (the "Software"), to deal in the Software

FIGURE 10-4:
The license information page.

The text of the license with your information filled in

6. **After you're satisfied with everything, click the Review and Submit button.**

   You return to the file creation page with the text of the license filled in.

7. **Scroll down to enter your commit message and click Commit New File to create the license file.**

Adding contributor guidelines

Another useful document to include in your repository is a file named CONTRIBUTING.md. This file provides information on how to contribute to your project. It should include information such as where people should ask questions and where to provide feedback.

Chapter 9 provides more details on what a contributor can expect to find in a CONTRIBUTING.md document. That chapter is a good place to start for your own project's contribution guidelines.

REMEMBER

A robust CONTRIBUTING.md document saves you and your future contributors a lot of headache and time. Be sure to revisit it from time to time to make sure it's up to date.

Adding a code of conduct

A code of conduct makes behavioral expectations clear for everyone who participates with your project, whether they contribute code, comment on an issue, or otherwise interact with others on your repository. A code of conduct sets the tone for the type of open and welcoming environment you want to foster for your project.

To add a code of conduct, follow the same process as you do when adding a license, but name the file CODE_OF_CONDUCT.md. (See the "Adding a license" section, earlier in this chapter.) This process is covered in Chapter 9 as well.

Making a Repository Public

At this point, your repository has the basic items it needs to go public. To change a repository from private to public:

1. **Visit the Settings page for your repository.**

2. **Scroll to the bottom of the Settings page to reach the Danger Zone, as shown in Figure 10-5.**

3. **Click the Make Public button to initiate making a repository public.**

WARNING

This step is a potentially dangerous operation because you may inadvertently expose information to the world you'd rather keep private. So make sure you really are ready to make this repository public. If you created it with the intention of making it public from the beginning, the repository shouldn't contain any secrets.

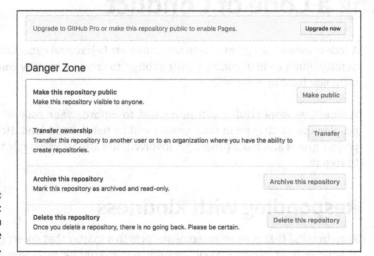

Upgrade to GitHub Pro or make this repository public to enable Pages.   **Upgrade now**

Danger Zone

Make this repository public
Make this repository visible to anyone.   **Make public**

Transfer ownership
Transfer this repository to another user or to an organization where you have the ability to create repositories.   **Transfer**

Archive this repository
Mark this repository as archived and read-only.   **Archive this repository**

Delete this repository
Once you delete a repository, there is no going back. Please be certain.   **Delete this repository**

FIGURE 10-5:
Danger zone: where you can make a private repo public.

GitHub displays a confirmation dialog after you click the Make Public button asking you to type the name of the repository to confirm this change, as shown in Figure 10-6.

4. **Type the repository name and then click the I Understand, Make This Repository Public button to make the repository public.**

   The Settings page now has a section that comes in handy as you manage your new open source project.

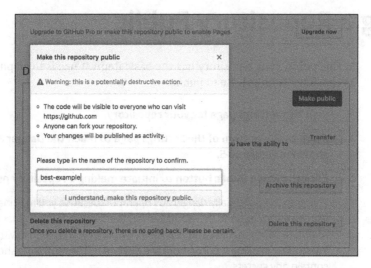

FIGURE 10-6:
Confirmation to convert a private repo to public.

Enforcing a Code of Conduct

A *code of conduct* is a great way to communicate behavioral expectations for a community. But a code of conduct isn't enough to create a healthy community on its own; you have to enforce it.

WARNING

It would be wonderful if you never had to enforce your code of conduct. Most people who participate in your project will be helpful and kind. However, as your project grows and more people get involved, it's almost inevitable that you'll have to step in.

Responding with kindness

Keep in mind that even when someone acts in a matter that doesn't fit in the spirit of your code of conduct, they may not necessarily be malicious or a bad actor. Maybe they're having a bad day, and they took it out on your repository. That doesn't excuse their behavior, but it is relatable. We've all done it.

TIP

If someone makes an off color or angry comment in an issue in your repository that goes against the type of community you want to foster, it's often disarming to reply back with a kind response that expresses empathy for their situation, but is clear that the comment doesn't meet community guidelines. Be specific in how it does not meet the guidelines and if you can, suggest an alternative approach they could have taken.

For example, if someone comments on the repository that the maintainer must hate other people because they haven't fixed a particular bug, you can respond with something like "Personal attacks are not welcome in this repository. We are all volunteers trying to work on this in our spare time. I understand that you are frustrated. Instead of a personal attack, perhaps write about how the bug affects you and why that is frustrating to you."

REMEMBER

But always remember, you don't owe anybody anything. You're not there to be a punching bag for people. If they're abusive or if you're simply tired of a thousand paper cuts of negative comments, rather than respond while angry, take a break. Walk away. Ask for help. Your own mental well-being comes first.

Leveraging the ban hammer

It would be nice if a kind admonishment worked in every situation (see preceding section), but sometimes tempers flare. We're all human, and sometimes we lose our temper. So will people who participate in your repository. This situation is where temporary interaction limits come in handy.

At the very bottom of the Settings page for your public repository (this setting doesn't exist for a private repository) is a moderation section with a link to Interaction Limits. Click this link to see the set of limits available. Figure 10-7 shows this page after enabling the Limit to Prior Contributors option for 1 week.

REMEMBER

These limits are not permanent bans. They are cool-down periods. While interaction limits across your repository remain in effect until you disable them, an individual restriction is intended to be a limit and not a ban. This tool is great to use when a discussion or an issue gets a bit too heated. It gives everyone time to cool down and prevents the discussion from running even further out of control. If a user who is not authorized to make changes to the repository attempts to make a change, they are met with a message. For example, if someone who hasn't contributed to a repository yet tries to open a pull request on a repository that has a limit to prior contributors, they see a message that states: "An owner of this repository has limited the ability to open a pull request to users who have contributed to this repository in the past.".

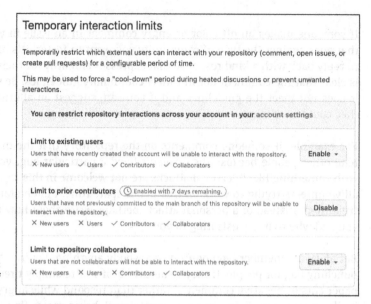

FIGURE 10-7:
The set of
interaction limits.

Blocking users

Every once in a while, you may encounter a truly abusive person in your project. You've tried to kill them with kindness. You've tried the 24-hour cool-down periods, but they persist in being rude or antagonistic. It's time to block or report the user.

To block or report the user:

1. Click the user's name next to their comment to visit their profile page.

Underneath their profile picture is a link to block or report the user, as shown in Figure 10-8.

2. Click the link to block or report the user.

A dialog box with two options appears: Block User or Report Abuse, as shown in Figure 10-9.

3. Click Block User to immediately block the user.

Blocking a user denies a user from accessing your activity and repositories. It also prevents them from sending you notifications, such as when they @ mention your username. They pretty much no longer exist as far as you are concerned, and vice versa. For more details on what happens when you block a user, visit https://docs.github.com/communities/maintaining-your-safety-on-github/blocking-a-user-from-your-personal-account.

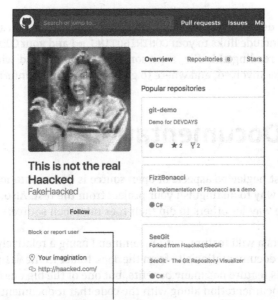

Link to block or report user

You can also block users, see a list of users you currently have blocked, and unblock users from your account Settings page by going to `https://github.com/settings/blocked_users`.

Writing a README.md File

The `README.md` file holds a special significance on GitHub. When you visit a repository's home page, GitHub displays the contents of the `README.md` rendered as Markdown on the home page. The `README.md` contents is what visitors to your repository expect to see to learn more about your project.

Chapter 3 includes information about what should be included in a README.md file. It should also include links to your CONTRIBUTING.md and your CODE_OF_CONDUCT.md files. After reading this file, a person should understand what your project does, how to get involved, and where to go to find out more information.

Writing Good Documentation

One of the most neglected aspects of open source is documentation. Writing good docs is a great way to distinguish your project from the rest. Also, documentation provides a nice way for others to dip their toes into open source.

TIP

GitHub supports a wiki feature, but I recommend using a relatively unknown feature, serving a documentation site from the docs folder in the main branch of your repository. This feature has many benefits, but one of the big ones is the ability to version your documentation along with the code that it documents.

Chapter 4 walks through building a GitHub Pages site. Serving your documentation from the docs folder is very similar to that process. The following steps assume that you have a repository that is not a GitHub Pages website. Chapter 3 walks through creating a regular repository if you need a refresher.

Make sure to commit a file named index.html to the docs folder in the main branch of your repository. This feature requires that there's already a docs folder before you can enable it. If you're unclear about how to do that, Chapter 7 covers writing and committing code in detail.

Navigate to your repository settings and then scroll down to the GitHub Pages section. By default, the Source is set to None. Click the button and select the main branch and then specify the /docs folder as shown in Figure 10-10 and then click the Save button.

You can now visit your documentation site at https://{USERNAME}.github.io/{REPO-NAME}/. To see an example of this in action, visit https://dra-sarah.github.io/best-example/.

As I cover in Chapter 4, you can select a theme and add a custom domain name if needed.

REMEMBER

Don't forget to add a link to the docs site to your README file so that people can find it.

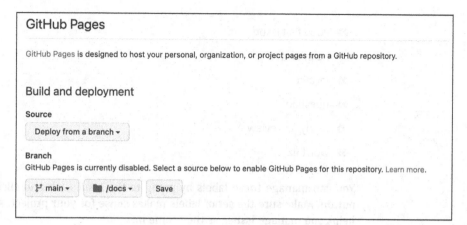

FIGURE 10-10:
Serving a GitHub
Pages site from
the docs folder.

Alternatively, you can deploy GitHub pages using a GitHub Action. Select GitHub Actions instead of Branch under Source, and choose one of the templates for Jekyll or static HTML.

TIP

If you want to still deploy only your docs subfolder, make sure you change the GitHub Action to point to docs` as the the path instead of `.`.

Managing Issues

As the word gets out about your open source project, people will start to try it out. And at some point, people will open issues. Congratulations! When an issue is opened, it means someone out there cared enough to report a bug, ask a question, or make a feature request. This interest is a good thing for an open source project.

But if your project becomes very successful, the influx of new issues can start to get overwhelming. In the following sections, I cover some tips on managing incoming issues.

Labeling issues

GitHub provides a default set of issue labels when you create a website.

» bug

» duplicate

» enhancement

- **»** good first issue
- **»** help wanted
- **»** invalid
- **»** question
- **»** ready for review
- **»** won't fix

You can manage these labels by going to the Issues tab and clicking the Labels button. Make sure the set of labels makes sense for your project. Assigning labels helps you manage issues as they come in.

Triaging issues

As new issues are created, it's important to triage issues. The word *triage* comes from the medical field. Merriam-Webster defines triage as "the sorting of patients (as in an emergency room) according to the urgency of their need for care."

Triaging issues is similar to triaging in a hospital. *Triage,* when applied to issues, is the process of reviewing new issues and assigning labels to indicate what type of issue are they (bug report, feature request, and so on), determining their priority, and assigning issues to people. In some cases, you may close issues you don't plan to address during triage.

TIP

Every triaged issue should have some label applied. If necessary, you can even create a triaged label for this purpose. Assigning a triaged label to issues that you've triaged makes it possible to see all issues that haven't yet been triaged by using the unlabeled filter, as shown in Figure 10-11.

> ☒ Clear current search query, filters, and sorts
>
> ☐ ⓘ **5 Open** ✓ 0 Closed   Author ▾   **Labels ▾**   Projects ▾   Milestones ▾   Assignee ▾
>
> ☐ ⓘ **Set up website alerts**   Filter by label
>   #5 opened 17 days ago by Haacked
>   Filter labels
> ☐ ⓘ **Do not use an alert message**
>   #4 opened 17 days ago by Haacked   Unlabeled
>
> ☐ ⓘ **Add a contribution section to F**   ■ bug
>   #3 opened 18 days ago by Haacked   Something isn't working
>
> ☐ ⓘ **The README should mention t**   ▪ duplicate
>   #2 opened 18 days ago by Haacked   This issue or pull request already exists
>
> ☐ ⓘ **Provide more details on the RE**

FIGURE 10-11:
Unlabeled
issues filter.

To maintain a sense of sanity, it often makes sense to schedule triage on a regular cadence as opposed to doing it all the time as people create new issues.

Issue templates

It would be great if everyone who submits an issue understood how to write a proper bug report or if they understood what makes for a good feature request. A bug report that just says "Your stuff is broken" is not too helpful to the repository maintainer.

To help foster good issues, you can set up issue templates that guide people filing issues on your repository to supply the information you need.

In the Settings page for your repository, in the Issues section in the main pane, is a big green button labeled Set Up Templates. Clicking that button takes you to a templates setup page where you can choose from a set of pre-existing issue templates or create your own, as shown in Figure 10-12.

FIGURE 10-12: Choosing an issue template.

From the templates setup page, follow these steps:

1. **Select both the Bug Report template and then select the Feature Request template.**

   You should see them listed on the page. Note they're not yet active. They still need to be committed to the repository. You can click the Preview and Edit button for each template if you want to change anything.

2. **Click the Propose Changes button to enter a commit message for these templates, as shown in Figure 10-13.**

3. **Click the Commit Changes button to commit these templates to the repository.**

They are stored in a special `.github/ISSUE_TEMPLATE` folder. If you navigate to that folder, you should see two files, `bug_report.md` and `feature_request.md`. Take a look at the contents of those files to understand the structure of an issue template.

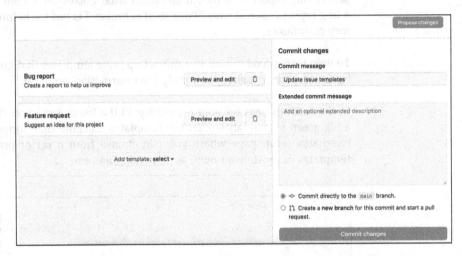

FIGURE 10-13: Choosing an issue template.

You can create new templates by going through the previous steps or by manually adding a file to the `.github/ISSUE_TEMPLATE` directory with the same basic structure as the existing issue template files.

Now, when someone creates an issue on your repository, they first see a set of issue templates, as shown in Figure 10-14.

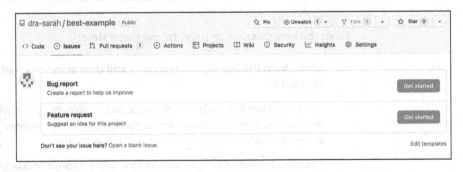

FIGURE 10-14: Choose an issue template to create an issue.

Clicking the Get Started button on one of the templates then displays the issue creation form, but with the contents prepopulated with the template information. Figure 10-15 shows an issue prepopulated with the default Bug Report template.

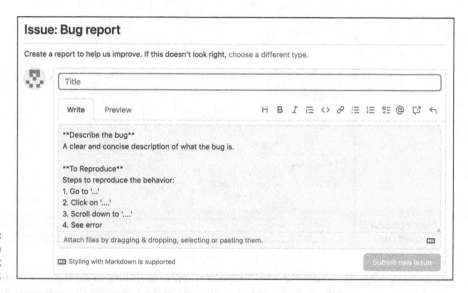

Issue: Bug report

Create a report to help us improve. If this doesn't look right, choose a different type.

Title

| Write | Preview |

H B I ≔ <> 𝒫 ≔ ≔ ≔ @ ☑ ↰

Describe the bug
A clear and concise description of what the bug is.

To Reproduce
Steps to reproduce the behavior:
1. Go to '...'
2. Click on '....'
3. Scroll down to '....'
4. See error

Attach files by dragging & dropping, selecting or pasting them.

ⓜ Styling with Markdown is supported

Submit new issue

FIGURE 10-15:
A new issue with
the Bug Report
template filled in.

The contents of the issue walks the issue creator through all the information expected of them. Of course, the person creating the issue can always choose to ignore the template and put anything they want in the issue.

Saved replies

Often, as your repository grows in popularity and the issues start to flow in, you'll find yourself repeating responses over and over again. For example, if someone reports a bug, you may need them to supply a log file. If every bug report should include a log file, then you should mention that in an issue template.

But even if you do, many folks will forget to include it. This situation is where a saved reply can come in handy. A *saved reply* is a canned response you can use over and over again.

As of this writing, canned responses apply only to a user and are not specific to a repository. So you can create your own saved replies, but they aren't immediately available to other members of your repository.

Go to `https://github.com/settings/replies` to manage your list of saved replies. To create a new saved reply, fill in the form at the bottom, as shown in Figure 10-16.

FIGURE 10-16:
Creating a new
saved reply.

To access the saved replies when responding to an issue or a pull request comment, click the arrow in the top-right corner of the comment box, as shown in Figure 10-17. That brings up a list of your saved replies. Click the one you want to use to fill the comment box with that reply.

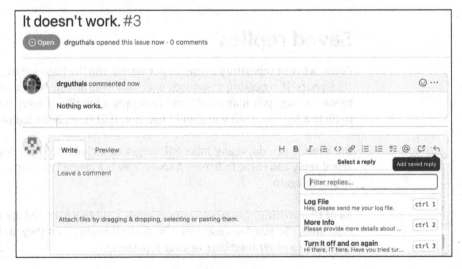

FIGURE 10-17:
Selecting a saved
reply in a
comment box.

Ending Your Project

At some point in the life of a repository, you may want to step away from it. This break may happen for many reasons. Perhaps the project is no longer useful, having been supplanted by other better projects. Perhaps the project is a runaway success, but you don't have the time to give it the attention it deserves.

Whatever the reason, it's important to handle the end of your involvement with the repository with the same care you showed in starting it.

Archiving a project

Archiving a project is a good option when a project is no longer all that useful to others. Or even if it's still useful, but mostly complete, archiving could be a good option. *Archiving* a project indicates that the project is no longer actively maintained. The code is still available to the world, and people can fork and star your project, but nobody can create new issues, pull requests, or push code to an archived repository.

To archive a repository, go to the repository settings page:

1. **Scroll all the way down to the Danger Zone and click Archive This Repository, as shown in Figure 10-18.**

    Clicking this button displays a detailed confirmation box that describes what will happen if you archive the repository, as shown in Figure 10-19.

2. **Type the name of the repository and click I Understand the Consequences, Archive This Repository to archive it.**

    Your repository is archived.

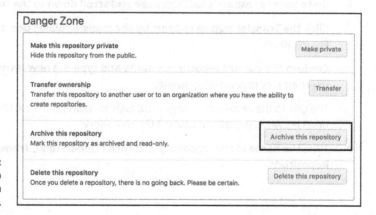

FIGURE 10-18:
The button to archive a repository.

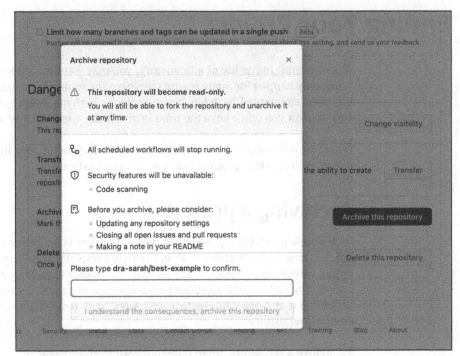

Transferring ownership

If your project is popular with a robust set of maintainers, you may want to transfer ownership to another account rather than archive the project. Transferring ownerships lets the new owner have full rights to the repository and continue to maintain it.

To transfer ownership:

1. **Go to your repository's Settings page and scroll down to the Danger Zone.**

2. **Click the Transfer button to bring up the transfer dialog box, shown in Figure 10-20.**

3. **Confirm the current repository's name and type the new owner's username or organization name.**

   The URL to the repository changes, but GitHub redirects the old URL to the new URL when anyone tries to visit the repository.

4. **Type the name of the repository and click I Understand, Transfer This Repository.**

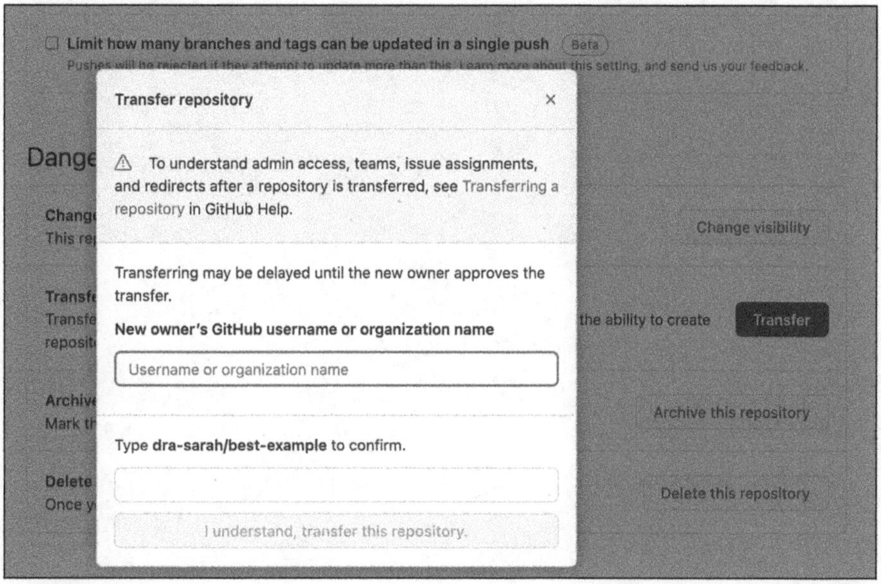

FIGURE 10-20:
The transfer repository dialog box.

Once you transfer the repository, you see a message at the top of the repository that says "Repository transfer to *<username>* requested," where *<username>* is the GitHub user or organization that you transferred the repository to.

The user you want to transfer the repository to then gets an email and can complete the transfer by clicking a link; if they do not accept the transfer in one day, the request expires. After clicking the link, they're taken to GitHub where they find a message at the top of the GitHub home page that says "Moving repository to *<repository-name>*. This may take a few minutes." Once the transfer is complete, the repository is removed from your account and added to the account of the person you transferred it to.

Chapter **11**

Inner-Source Your Code on GitHub

Chapters 9 and 10 talk about the open source community and best practices on GitHub.com. But if you want to keep your code private, GitHub offers unlimited private repositories for free, personal plans to support your private software development.

In this chapter, you discover some situations where you may want to code in private. You also get the inside scoop on inner-sourcing your code.

Why Code in Private?

If you're appropriately representing the open source community, you may be inclined to do everything in the public. But what if you work at a company on proprietary code? What if you're starting a new company where you plan on selling software? Or, what if you're a student, and you're working on a group coding assignment? These are examples of when it may be appropriate to work in a private, collaborative environment, when it may be appropriate to inner-source your code.

Inner-sourcing is really just a play on words with open source. It implies that you use the same (or similar) strategies to collaborative code writing as open source, but you do it on private repositories.

Using GitHub Organizations

Depending on whom you're working with and the scope of the project, you may want to have access to certain GitHub features. If you're working on proprietary code for a company (or code that may be a product for a future company you're hoping to start), then the investment of GitHub Teams may be suitable, giving you private repos as part of an organization. But if you're working with other students at your university on a semester-long project, then just creating a private repo may be good enough.

Creating a GitHub organization

GitHub *organizations* allow you to give a group of users access to a set of repositories all at once. With features such as *teams*, you can also have subgroups of users who have different access rights to different repositories. This setup also makes communicating across GitHub easier because you can tag an organization (or team) instead of an individual person. If you're running a large project that has more than one repository, organizations may be a good option for you.

You can start an organization in one of three tiers:

>> **Free:** In this tier, you get unlimited public and private repositories, unlimited collaborators, issues and bug tracking, and project management. Essentially, this tier allows a core group of people who are working on an open source project to work together easier. If you don't need advanced features such as GitHub Codespaces or workflow features on private repositories such as code owners or protected branches associated with the organization, this tier is a great option.

>> **Team:** In this tier, you get everything from the previous tier, plus 32 GB of GitHub Codespaces, advanced code workflow features on private repositories, pages and wikis on private repositories, and advanced insights into all your repositories.

TIP

Two-factor authentication is available on all organizational tiers and is when you're required to enter your password plus perform an additional security measure to ensure it is really you logging in to the website. Because passwords can sometimes be hacked, organizations often require that you use a mobile app (such as Duo Mobile or Google Authenticator), text-message confirmation, or a physical YubiKey to verify that it is you.

>> **Enterprise:** In this tier, you get everything from the previous two tiers, plus more CI/CD minutes per month, more package storage, and GitHub Connect, which allows you to share features and workflows between your GitHub Enterprise Server instance and GitHub Enterprise Cloud.

REMEMBER

Continuous Integration/Continous Deployment CI/CD is a category of developer tools that use automation to consistently merge small changes into your main branch (integration) and consistently deploy changes from the main branch to production (deployment).

You can find the tiers by clicking the plus symbol at the top right of GitHub.com (when you're logged in) and choosing New Organization. You see a new page where you can choose the organization's name, specify a billing email address, and choose a tier.

Don't worry: If you choose the free tier, you aren't billed anything; you won't even be asked for a credit card.

Inviting members to your GitHub organization

During the organization setup process, you're asked if you want to invite others to join your organization. Using their GitHub.com alias, you can search for them in the provided box and send them an invite. If you forget to invite someone or want to invite them later, you can still do so by going to the People tab on the organization's home page and clicking Invite Member. You can choose the permission level for the person you invite to your organization, shown in Figure 11-1. You can change the permission level later, if you need to.

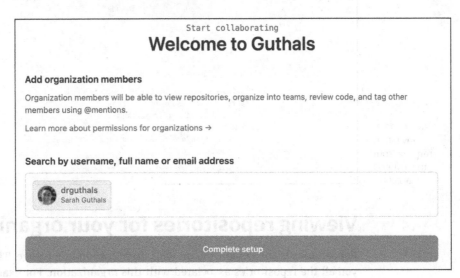

FIGURE 11-1: Inviting someone to join your organization.

If you're on the free tier, you have unlimited seats, but if you're on the paid tier, then you have only the number of seats you're paying for.

TIP

The Teams tier has a minimum of five seats at a set price. If you want to add more seats, you can, but you always start with five.

After you invite someone to your organization, they receive an email notification with a link to accept the invitation, or they can go to the organization's GitHub home page to accept the invitation. For example, in Figure 11-1, I invited Sarah to be a part of the Guthals organization because it's my primary account. When I visited `https://github.com/Guthals` while logged in to my primary GitHub account, I saw a banner and View Invitation button at the top of the page. Clicking the View Invitation button, I was taken to a new page, shown in Figure 11-2, where I was able to first see what kind of access the owners of the Guthals organization would have to my primary GitHub account information if I were to join.

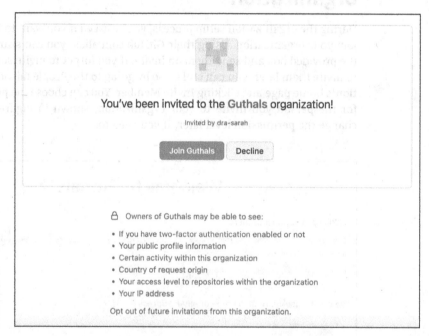

You've been invited to the Guthals organization!

Invited by dra-sarah

Join Guthals   Decline

🔒 Owners of Guthals may be able to see:

- If you have two-factor authentication enabled or not
- Your public profile information
- Certain activity within this organization
- Country of request origin
- Your access level to repositories within the organization
- Your IP address

Opt out of future invitations from this organization.

FIGURE 11-2: Invitation request from the invitee's account.

Viewing repositories for your organization

Repositories is the default tab on your organization home page. This page shows you all the repositories associated with this organization. For example, if you go to the Microsoft organization home page on GitHub (`https://github.com/microsoft`), you find more than 5,000 repositories and more than 4,000 people involved in open source projects for Microsoft (see Figure 11-3).

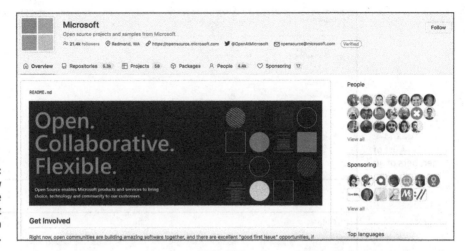

FIGURE 11-3:
The overview
page for the
Microsoft
organization on
GitHub.com.

The Overview page has a `README.md` file at the top that describes how to get involved with open source projects maintained by the Microsoft organization. Pinned repositories appear just under the organization's `README.md`. Pinned repositories are the ones that the Microsoft organization owners think are the most relevant to folks interested in what Microsoft is doing in the open source space. For example, the VS Code repository has more than 140,000 stars and more than 24,300 forks. As one of the most popular editors and most popular open source projects, Microsoft wants to make sure this repository is front and center for visitors to their open source organization home page.

Managing members of your organization

You will always have at least one member of your organization — you! But this section is more interesting if you have more than one member, so if you haven't invited other members yet, go to the earlier section "Inviting members to your GitHub organization" and invite someone else.

To see all your organization's members, from your organization home page, click the People tab. You should see all the members of your organization on this tab, as shown in Figure 11-4.

TIP

On the right of each member is a small cog drop-down menu. This menu gives you quick options for managing your organization members. You can also get all these options and more information about a specific member by choosing Manage from that drop-down menu. You then see an overview of an organization member, as shown in Figure 11-5.

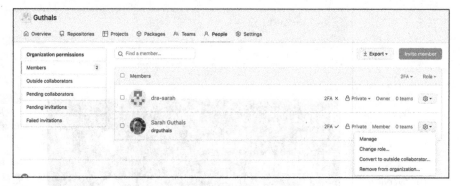

FIGURE 11-4:
A list of
members of an
organization.

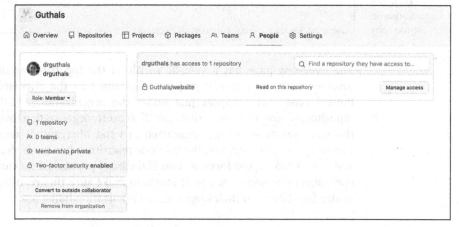

FIGURE 11-5:
An overview of a
member of an
organization.

Figure 11-5 gives you the following information about the person:

>> **Role:** The person's role in the organization with the ability to change it to a different role from this page.

>> **Repository access:** The number of repositories this person has access to within the organization, as well as a list of all the repositories and what permissions they have for each one. Each repository has a button that allows you to quickly navigate to the settings for that person for that repository.

>> **Number of teams:** The number of teams the person is a part of within the organization.

>> **Activity:** Information on whether the person is sharing their activity on projects within this organization on their public profile.

>> **Two-factor authentication:** Whether this person has two-factor authentication enabled for their account. Two-factor authentication can be a requirement of your organization, which you can change in the settings for your organization. (See the upcoming section "Setting organization settings.")

» **Convert to outside collaborator:** A button to convert someone to an outside collaborator. This feature is useful for short-term or very scoped projects. Instead of having a person be a part of the entire organization, you can make them a part of a single team that has access to certain repositories, making their privileges easier to manage.

» **Remove from organization:** As straightforward as it sounds. This setting removes the person from the organization.

Creating teams within your organization

As your organization begins to grow, it may make sense for you to create teams within your organization. The benefit of teams is that you can quickly give access to a repository to an entire team, without having to remember every single person that is on that team. To create a team, click the Teams tab on your organization home page and click New Team. A new page appears where you can choose a team name, add a description, choose a parent team (if you've created other teams already), and set the team's visibility within the organization.

REMEMBER

Choosing a team name is an important part to consider carefully. The team name is how folks within your organization can tag everyone in the team all at once. For example, if you have a security-vulnerabilities team that manages all security vulnerabilities for your website and a security bug is found, you can tag the security-vulnerabilities team, and each person on that team will get a notification. This will help make sure that you get the fastest response time, accounting for different time zones, working hours, shifts, and schedules. You want the name to be representative of the team because if you had named the team something like powerpuffgirls, it would be pretty confusing to see an issue comment that says

```
@powerpuffgirls, please review this security vulnerability asap
```

Creating teams has a lot more benefits than the ability to mention them all using one alias. If you're interested, read the section "Making the Most of Your Teams," later in this chapter.

Setting organization settings

Organizations have a few more settings than typical individual GitHub accounts. On the right-most Settings tab on the organization home page a Settings page that is similar to the one I describe in Chapter 1, but with some key differences:

» **General** is where you can change the organization's name, avatar, and primary contact email address or even delete the organization. You can also choose to join the GitHub Developers Program (you can read about at `https://developer.github.com/program`). If your organization is representing a corporation, you can even sign the corporate terms of service, which helps protect your IP even on public repositories.

>> **Access** is where you can specify what level of access your members and repositories have. Here you find the billing and plans, repository roles, member privileges, team discussion settings, functionality to import and export data, and options for moderation.

TIP

Under the billing and plans setting, you can add billing managers to your organization; this can be really useful for folks within your company who need to have access to billing information, but may not be savvy or interested in the code aspect.

>> **Code, planning, and automation** is where you can specify settings across your repositories, manage your Actions, webhooks, discussions, packages, pages, and projects.

>> **Security** is the area where you can require that all members of your organization have two-factor authentication, set up SSH certificates, create IP allow lists, manage code security, and verify a domain that you own so that you can verify your organization's identity on GitHub.

>> **Third-party access** is all third-party applications that you have given access to your repositories.

>> **Integrations** has the integrations with your organization; you can connect a Slack organization to your GitHub organizations to get started.

>> **Archives** has the log of all activity done to the organization (not the individual repositories part of the organization), and a list of deleted repositories.

>> **Developer Settings** has two settings — OAuth Apps and GitHub Apps — and a place to specify management of the organization (under GitHub Apps).

TIP

Moderation settings, under the Access menu, is where you can block users from your organization or enact temporary (24-hour) limitations on what activity can happen on any repository within the organization. Figure 11-6 shows that you can block users for short periods of time or forever. These moderation settings are often most useful in open source project organizations because each member is less likely to have a common driving force, such as a paycheck, to behave with respect. However, moderation settings can also be useful with inner-source projects if things simply start to get heated within the organization.

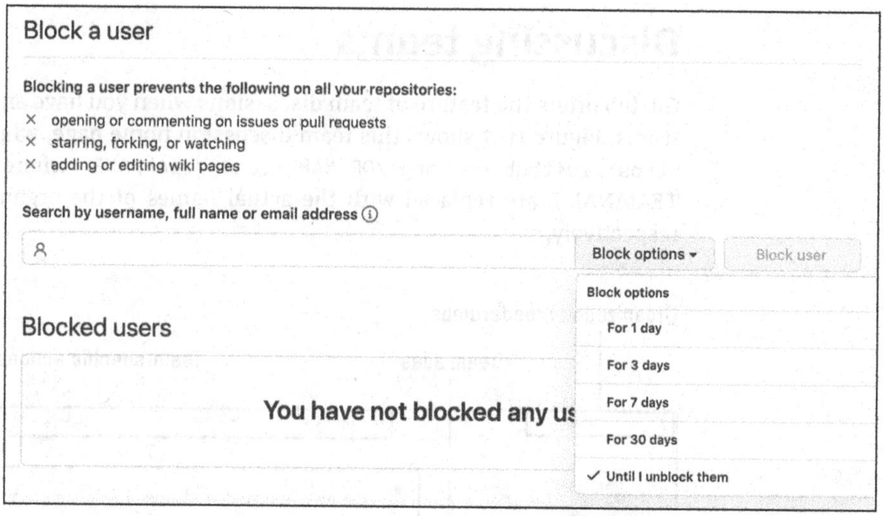

Block a user

Blocking a user prevents the following on all your repositories:

× opening or commenting on issues or pull requests
× starring, forking, or watching
× adding or editing wiki pages

Search by username, full name or email address ⓘ

Block options ▾   Block user

Block options

For 1 day

For 3 days

Blocked users   For 7 days

For 30 days

You have not blocked any us   ✓ Until I unblock them

FIGURE 11-6:
Options for
blocking an
individual.

Making the Most of Your Teams

Having an organization with members grouped into various teams can be useful simply for understanding who is doing what, but GitHub provides you with tools that make your workflow even more effective when you group organization members into teams.

Creating parent/child teams

You can add a team as a parent team to a new team that you're creating (see the section "Creating teams within your organization," earlier in this chapter). Essentially, teams can have *hierarchies*. The permissions of the parent teams are passed down to all child teams, but not vice versa. Hierarchies can become extremely useful when you want to ensure that everyone within an organization has access to exactly what they need and no more, no less.

As an example, you may have a team named Employees at the top of your hierarchy. As part of that team, you may have three other teams: Human Resources, Marketing, and Engineering. Under the Human Resources team, you may have some private project boards or private repositories that only the Human Resources team members should be able to see (such as employee personal information). Under the Marketing team, you may have customer data that shouldn't be shared publicly within the organization (such as billing account information). By assigning all the employees to respective child teams within the organization, you can ensure that everyone has access to the employee handbook, while only the folks in Human Resources have access to Social Security numbers.

Discussing teams

GitHub offers the feature of team discussions when you have an organization with teams. Figure 11-7 shows this team discussion home page, which you can find at `https://github.com/orgs/ORGNAME/teams/TEAMNAME` where ORGNAME and TEAMNAME are replaced with the actual names of the organization and team, respectively.

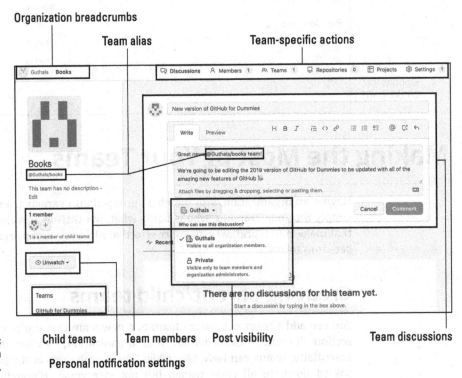

FIGURE 11-7: Discussions on a Team page.

From the team discussion home page, you find

>> **Organization breadcrumbs:** On the top left, the hierarchy of this team, all the way up to the organization in a clickable breadcrumb-like fashion for easy navigation.

>> **Team-specific actions:** On the top right, similar menu items that are found on repositories and the organization itself, but all are specific to this team.

>> **Team discussion:** In the center right, a place to add new posts to the team discussion page. Below this area, you can find previous posts, and you can even pin specific posts to stay at the top (useful for important announcements or onboarding information).

>> **Team alias:** The team alias appears below the team avatar and name on the left-hand column. You can use this alias to notify everyone in the team about something that may be of interest to them, as seen in the team post.

>> **Team members:** A list of all team members, a count of how many also belong to child teams, and a button for adding additional members to the team.

>> **Personal notification settings:** On the left-hand column, a place to quickly change your notification settings for this team.

>> **Child teams:** A list of all child teams to this team that are linked for quick navigation, located on the bottom of the left column.

Assigning code owners

A really neat feature that GitHub has that is enhanced with the use of teams is the CODEOWNERS file. Code owners is a way to specify certain people who may have the most experience and knowledge on a piece of code. They're the people who, when you're looking for someone to review your code, you most likely want to take a look. They probably have the most history with the code or are the most expert in that domain. By creating teams within your organization, you can assign multiple people to be code owners of certain code, without having to assign each person or keep the list up to date.

To get code owners working on one of your repositories, first make sure that you have a team setup that includes members of the organization who should be held responsible for ensuring all code that gets added to that repository is correct. In my example, I have a team called GitHub For Dummies. On that team page, make sure that you've linked the repository that you want this team to be code owners of and make sure that this team has at least Write access to that repository. Figure 11-8 shows that my team has two repositories that I have Write access to: the public one that you can access as a part of this book, and a private one that I'm using to track the progress of writing this book. I'm using code owners for the public repository.

FIGURE 11-8: Repository access for a specific team.

WARNING

If you give the team only Read access to the repository, then they can't approve or merge a pull request because that is technically writing to the repository. Code owners (whether individuals or teams) must have either Write or Admin access to be effective.

In the repository where you want the code owners to be automatically added as a reviewer for all pull requests, add a file on the main branch .github/CODEOWNERS with the following code:

```
# These owners will be the default owners for everything in
# the repo. Unless a later match takes precedence,
# @thewecanzone/github-for-dummies will be requested for
# review when someone opens a pull request.
* @thewecanzone/github-for-dummies
```

Make sure under the Settings for that repository, you've added a branch protection to ensure that every time someone tries to make changes to the main branch using a pull request, they have to get their code reviewed and approved by at least one person and that each pull request requires a code owner to review and approve it. You can set up these requirements in the Branches area of the Settings for the repository, as shown in Figure 11-9.

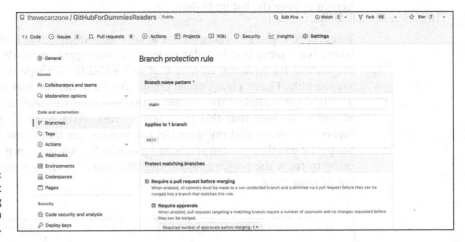

FIGURE 11-9: Requiring at least one approving review from a code owner.

Now, when you try to add code by opening a pull request, the team @thewecanzone/github-for-dummies is automatically added to every single pull request, as shown in Figure 11-10.

FIGURE 11-10:
Code owners are automatically added to a pull request.

TIP

You can also add specific people or more teams to the CODEOWNERS file by appending them to the same line. And you can even specify what type of files you want to assign to which code owners. For a detailed explanation on all of the nuances, you can visit `https://help.github.com/articles/about-code-owners`.

Best Practices for Inner-Sourcing

In Chapters 9 and 10, I describe best practices for both contributing and creating your own open source project. Here is the secret: Those best practices all apply to inner-sourcing as well. Even if you're working on your code in private on a small team, it's still important to document, follow style guidelines, communicate effectively through issues and pull requests, provide applicable feedback through thorough code reviews, and be a positive team member.

REMEMBER

When in doubt, pretend the world can see you.

In addition to following best practices for open source on your private projects, you can also leverage additional GitHub resources because your projects are within a private organization.

Repository insights

When you're a part of a company, evaluating an engineer on their contributions to the team can get tricky. Either as an individual trying to make sure you're making good progress or as a manager trying to write end-of-year reviews and performance ratings, you want to avoid using information to qualify someone's overall contributions and performance.

That being said, you can get some information to help guide conversations to more effectively evaluate yourself or another individual. On a repository, you can click the Insights tab at the top right. You see a list of insights about your repository and the people that contribute to it:

>> **Pulse** is a snapshot of your repository for a specific time period (defaulted to one week). For example, in the month of October 2022, the Microsoft/VSCode project had the following pulse summary:

```
Excluding merges, 36 authors have pushed 208 commits to main
    and 319 commits to all branches. On main, 481 files have
    changed and there have been 10,619 additions and 7,883
    deletions.
```

>> **Contributors** shows an overall contribution graph for the repository, as well as individual contribution graphs for each contributor with a summary of the number commits, lines added, and lines removed.

WARNING

Tracking contributions to a project simply by lines of code modified or number of commits isn't an effective way to evaluate someone if nothing else is taken into account. For example, Shawn could commit every 15 minutes for fear of losing work, and Sam could always focus his energies on refactoring, making the number of lines of code they change substantial. Comparing these two to Sandra, who had only a few commits and not as many lines of modified code but found and fixed a security vulnerability that could have taken down the application for good, isn't a fair comparison.

>> **Community/Traffic:** On public repositories, the community profile on a repository helps maintainers understand where they can improve their repository to better support their community. You can find out more at https://help.github.com/articles/about-community-profiles-for-public-repositories. On private repositories, the Traffic section gives you insight into who is coming to this repository, where they're coming from, and what files they're interacting with.

>> **Commits:** This simple commit graph can give you insights into problems your repository may have or trends with the lives of your contributors. For example, if contributions drop off every year around May, maybe a computer science student contributes to your repository as a part of the course they are taking. And if you typically have 300 commits a week but suddenly you start getting only 50 commits, maybe folks are running into a problem where they can't get their code working well enough to commit.

>> **Code frequency:** Similar to commits, code frequency shows the frequency at which lines of code are added and lines of code are deleted. Identifying patterns here can help you understand the health of your code as well as the profile of your contributors.

WARNING

>> **Dependency graph:** The dependency graph lists all dependencies (and dependents) of this repository and the version it depends on.

If you want GitHub to create a dependency graph for your private project and therefore be able to warn you if you have any dependency security vulnerabilities, you have to grant GitHub read-only access to your private repo (see Figure 11-11). (For more on dependency security vulnerabilities, read https://github.blog/2018-07-12-security-vulnerability-alerts-for-python/ for the announcement on security vulnerability alerts for Python.)

>> **Alerts:** On a private repository, you have an Alerts section. If you've enabled read-only access to GitHub either via allowing it through the dependency graph setting shown in Figure 11-11 or in the Settings tab for the repository in the Options category under Data Services, as shown in Figure 11-12, you see security alerts here.

>> **Network:** The network graph shows all the people who have branched and forked the repository and any branches of branches or forks of forks. Essentially, it shows all the possible states of your repository in the world today. Figure 11-13 shows the network graph for Visual Studio Code.

>> **Forks:** The Forks tab lists links to all forks of your repository.

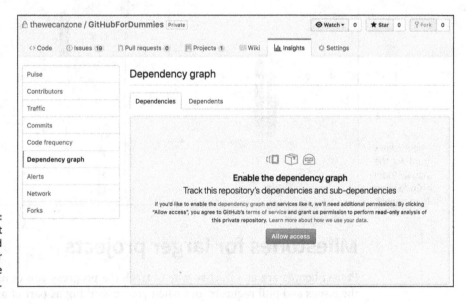

FIGURE 11-11: GitHub must have read access to your repository to give security alerts.

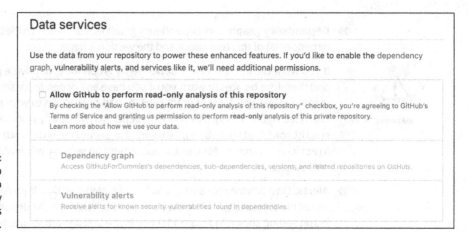

Data services

Use the data from your repository to power these enhanced features. If you'd like to enable the dependency graph, vulnerability alerts, and services like it, we'll need additional permissions.

☐ **Allow GitHub to perform read-only analysis of this repository**
By checking the "Allow GitHub to perform read-only analysis of this repository" checkbox, you're agreeing to GitHub's Terms of Service and granting us permission to perform **read-only** analysis of this private repository. Learn more about how we use your data.

☐ Dependency graph
Access GitHubForDummies's dependencies, sub-dependencies, versions, and related repositories on GitHub.

☐ Vulnerability alerts
Receive alerts for known security vulnerabilities found in dependencies.

FIGURE 11-12: Allow GitHub read access to your repository for various data services.

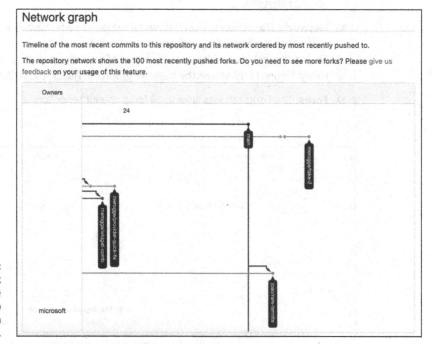

Network graph

Timeline of the most recent commits to this repository and its network ordered by most recently pushed to.

The repository network shows the 100 most recently pushed forks. Do you need to see more forks? Please give us feedback on your usage of this feature.

FIGURE 11-13: The network graph for the Visual Studio Code open source project.

Milestones for larger projects

Project boards are an effective way to track the progress you're making through the issues and pull requests, but when you're working as part of a larger organization, you often have larger milestones that you're trying to reach. GitHub provides support for milestones that can be linked to issues and pull requests.

To create a milestone:

1. **Go to the Issues tab of your repository and click the Milestones button.**

If you don't have a milestone yet, you can click the big, green New milestone button on the top right or the big, green Create a Milestone button in the center.

2. **Add a title and optional due date and optional description.**

3. **Click Create Milestone.**

You see a list of milestones.

In your list of milestones, to the right of the one you just created, is a progress bar with a status of the percentage of issues and pull requests completed, the total number of issues and pull requests still open, and the total number of issues and pull requests already closed. This information can give you a quick snapshot of how close you are (or, in most cases, aren't) to meeting your deadline.

On your Issues or Pull Requests tabs, you can add the milestone to any issue or pull request, and the status icon on the milestone list updates automatically. Clicking the milestone gives you a list of all the issues and pull requests associated with that particular milestone.

5

Making GitHub Work for You

Chapter **12**

Collaborating Outside of GitHub

W hile GitHub may be the hub of a software development project (it hosts the key deliverable, the source code), it's not the only place where collaboration occurs. Software development teams use a variety of tools to communicate and coordinate their software efforts. Many people who are not developers also work on a software project and need to be kept apprised of the progress of a project in the tools they use.

For example, a lot of day-to-day collaboration occurs in chat rooms, such as Slack. Others may use a Trello board to manage tasks for a team. Still others may use Octobox to keep on top of their GitHub notifications.

In this chapter, I explore the various integrations that bring GitHub information into other collaboration tools. This chapter is in contrast to Chapter 13 where I cover integrations that bring information from other tools into GitHub to improve the software development workflow.

Chatting It Up

For many teams, especially distributed teams, chat is a powerful way for members of the team to collaborate and coordinate their efforts. Chat in this context does not refer to sipping tea on a porch talking about how their day went. *Chat* refers to text-based tools, such as Slack, used by teams to communicate both synchronously and asynchronously.

Many teams find it helpful to have GitHub post important notifications into a chat room so teams are kept apprised of what's going on with a repository. In this section, I set up a GitHub integration with one popular chat software, Slack.

Before you install the integration, you need to be the admin of a Slack workspace. You can create a free Slack workspace at https://slack.com.

After you set up your Slack workspace, installing the GitHub for Slack integration requires two key steps:

1. Install the GitHub app for Slack in the Slack workspace.

2. Add the Slack App for GitHub to your GitHub account.

The following sections cover these steps in detail.

Installing the GitHub app for Slack

To install the GitHub app for Slack:

1. Go to https://slack.github.com and click the Add to Slack button in the center of the browser window.

   If you're not logged into your Slack workspace in the browser, clicking the Add to Slack button prompts you to sign into your slack workspace. Likewise, if you're not logged into GitHub, the site prompts you to log into GitHub. When you're authenticated to both, you see a Slack confirmation screen with information on what permissions the GitHub app has to your Slack workspace. Figure 12-1 shows the confirmation screen with every section expanded (they're collapsed by default), so you can see everything the integration can do.

   TIP

   Be sure you're adding the GitHub app to the correct Slack workspace. If the wrong workspace is shown, you can change it by clicking the workspace drop-down list on the top right of this screen. Here, you can choose to change which Slack workplace you want to install the integration into. If the workspace isn't listed there, you can click Add Another Workspace.

2. **Click Allow, shown in Figure 12-1, to be taken to Slack to authenticate with GitHub.**

   As shown in Figure 12-2, this DM prompts you to sign in with GitHub.

3. **Click Connect to GitHub Account, and then enter the post authentication code.**

4. **Check out the instructions for how use the GitHub Slack integration.**

   As shown in Figure 12-3, you can subscribe specific Slack channels to specific repositories, create issues, and manage notifications from inside Slack.

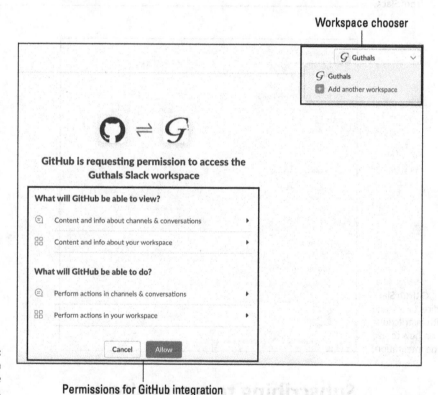

FIGURE 12-1:
Confirmation
page for the
GitHub app.

WARNING

If you have the Slack desktop application, clicking the Install button attempts to launch the application and take you to the workspace where you installed the app. If you haven't yet added the workspace to the desktop application, you'll just be in the last workspace you used. This can be a bit confusing as it may seem like the installation didn't work. Don't worry; it probably did work. Just add the workspace to your desktop application and continue.

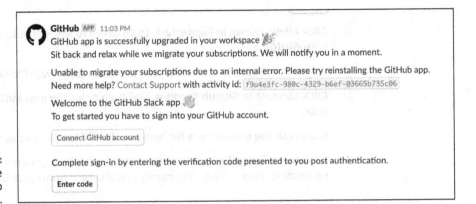

FIGURE 12-2:
Authenticate
with GitHub
from Slack.

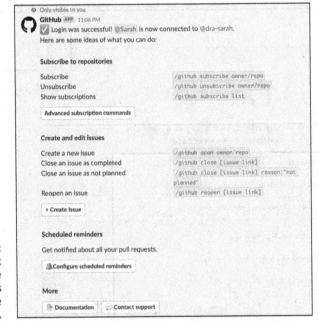

FIGURE 12-3:
GitHub Slack
direct message
with instructions
on how to use
the integration.

Subscribing to a repository in a Slack channel

After you install the GitHub app for Slack, you can subscribe to notifications for a GitHub repository from within a Slack channel, by typing

```
/github subscribe owner/repository
```

For example, to subscribe to the repository I created for the readers of this book, `https://github.com/thewecanzone/GitHubForDummiesReaders`, you would type the following in a Slack channel:

```
/github subscribe TheWecanZone/GitHubForDummiesReaders
```

When you've successfully subscribed to a specific repository, you see the confirmation message shown in Figure 12-4.

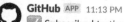

GitHub APP 11:13 PM
☑ Subscribed to thewecanzone/GitHubForDummiesReaders. This channel will receive notifications for

`issues`, `pulls`, `commits`, `releases`, `deployments`

Learn More

TIP

If you want to know more about how GitHub works with Slack, click the Learn More link or go to `https://github.com/integrations/slack#configuration` to check out the open source repository for the GitHub Slack integration.

Trying out the GitHub Slack integration

With the installation complete, you can now subscribe to GitHub repositories in your Slack channels. To see the full list of Slack commands, type the following:

```
/github help
```

The output of this command is the same output you get when you first install the integration (refer to Figure 12-3).

To test the GitHub app and open a new issue:

1. **Run the following command:**

   ```
   /github open TheWeCanZone/GitHubForDummiesReaders
   ```

   A Slack dialog box appears. You can use this dialog box to create a new issue, as shown in Figure 12-5.

2. **Fill in the dialog box and click the Open button.**

   Clicking the Open button creates the issue on GitHub. And because I'm subscribed to that repository, I get a Slack message in the channel that the issue was created, as shown in Figure 12-6.

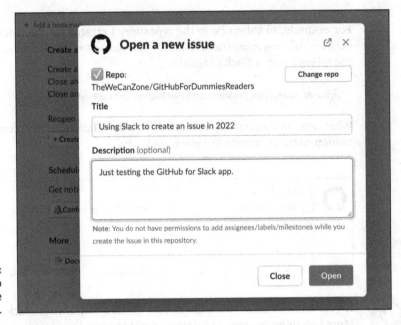

FIGURE 12-5:
Slack dialog to
create an issue
on GitHub.

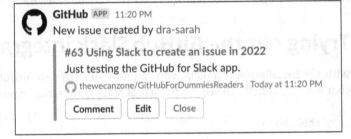

FIGURE 12-6:
Slack message
with information
about a newly
created
GitHub issue.

TIP

If you find a bug with the GitHub integration or have an idea for a way it could be better, good news! It's open source on, of course, GitHub! You can log issues or even contribute at `https://github.com/integrations/slack`.

The `/github subscribe` command by default subscribes a channel to notifications for the following features of a repository:

» `issues`: Opened or closed issues

» `pulls`: New or merged pull requests

» `commits`: New commits on the default branch (usually `main`)

» `releases`: Published releases

>> `deployments`: Updated status on deployments

>> `reviews`: New reviews completed on pull requests

You can remove a single feature by using the `/github unsubscribe owner/repo [feature]` command. For example, to remove commit notifications on the default branch, run the following command.

```
/github unsubscribe TheWeCanZone/GitHubForDummiesReaders commits
```

Getting Trello and GitHub Integrated

Trello is a collaboration tool used to organize projects into boards, lists, and cards. It's inspired by the Kanban scheduling system popularized by Toyota. *Kanban* is Japanese for signboard. The idea is to have a board that provides a view of a project's status and progress at a glance.

Often, a tool like Trello is combined with GitHub to manage a project. A project team may use Trello to manage the entire project, but use GitHub to host the code and assign specific code issues to developers. A card in Trello might correspond to multiple GitHub issues.

TIP

GitHub project boards are essentially Trello with GitHub already integrated into it. However, project boards within GitHub aren't always the right fit for your team. Some teams have nondeveloper folks who don't want to have to learn GitHub and may already be using Trello. Typically, it's best to have all your project management in one place, so if that should be outside of GitHub, in Trello for example, you can still make it a part of your developer workflow with this integration.

A GitHub integration (what Trello calls a *power-up*) for Trello connects cards to GitHub issues, pull requests, and branches. In the next section, I walk through setting up the GitHub power-up.

Installing the GitHub power-up

The following installation instructions assume that you've already signed up for `https://trello.com` and created a project board:

TIP

If you've never used Trello, you can visit its guides at `https://trello.com/guide`. It is similar to GitHub project boards (see Chapter 3). For specific help on creating a board and cards, visit `https://trello.com/guide/create-a-board.html`.

1. **With a board open, make sure the menu is open.**

   If not, click in the top right to show the menu.

2. **Click the Power-Ups section of the menu, as shown in Figure 12-7.**

   Clicking the Power-Ups button brings up a search dialog box for power-ups.

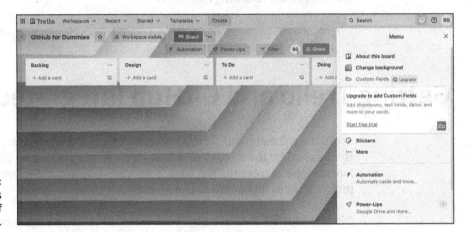

FIGURE 12-7:
The power-ups
section of
the menu.

3. **Search for GitHub to find GitHub related power-ups.**

4. **Click the Add button for the GitHub power-up to enable it.**

TIP

   Just like for any application, Trello may have GitHub power-ups (extensions/integrations) that are built by GitHub and some that are built by other folks. Because GitHub's API is public, folks can often create power-ups/extensions of their own. Be sure you're always aware of the author of the power-up/extension when you're installing it. You may very well want to install from a third-party developer instead of GitHub itself because the features might be different. Regardless, you should make sure you're aware of that choice. Don't assume anything with "GitHub" in the title is made by GitHub the company.

5. **After you enable the power-up, click the settings button to configure it.**

   You see a menu with the option to authorize or disable the power-up.

6. **Click Authorize Account.**

   An option to link your GitHub account appears.

7. **Click Link Your GitHub Account.**

   GitHub.com launches in your browser and prompts you to Authorize Trello, as shown in Figure 12-8.

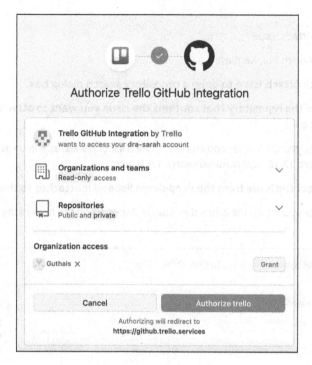

FIGURE 12-8:
Authorize Trello
on GitHub.

8. **Click the Grant button next to any organizations that you want to connect with Trello.**

   In my case, I granted Trello access to the guthals organization.

9. **Click the Authorize Trello button to make the power-up active.**

Using the GitHub power-up

The GitHub power-up is accessed via the power-up button on the back of any Trello card. If you haven't already, go ahead and create a couple of cards.

To use the GitHub power-up on your Trello board, follow these steps:

1. **Click the card to access the back of the card.**

   Figure 12-9 shows a card that I created. The GitHub power-up shows up in the bottom-right corner.

2. **Click the GitHub Power-Up button.**

   Four menu options appear:

   - Attach Branch
   - Attach Commit

- Attach Issue

- Attach Pull Request

3. **Click Attach Issue to open a repository search dialog box.**

4. **Find the repository that contains the issue you want to attach to the card.**

   After you select the repository, you see a list of issues, as shown in Figure 12-10. You can also search for issues.

5. **Select the issue from the drop-down list and it attaches to the Trello card.**

   After you attach the issue, the issue is displayed on the back of the Trello card.

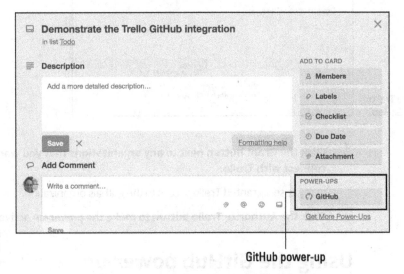

FIGURE 12-9:
Trello card with a
GitHub power-up.

GitHub power-up

A Trello card may be attached to multiple GitHub items. For example, repeat the previous steps, but choose Attach Pull Request instead of Attach Issue to attach a pull request to an issue. When you are done, you see both an issue and a pull request attached to the Trello card, as shown in Figure 12-11.

The front of the card shows a couple icons that indicate that this card is attached to GitHub. It shows an Octocat icon with a count of GitHub items attached to the card. It also shows pull request icon with a count to indicate the number of pull requests attached, as shown in Figure 12-12.

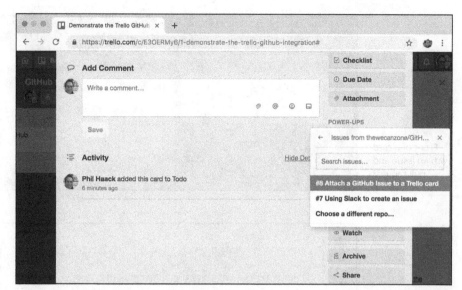

FIGURE 12-10:
Selecting the
issue to attach.

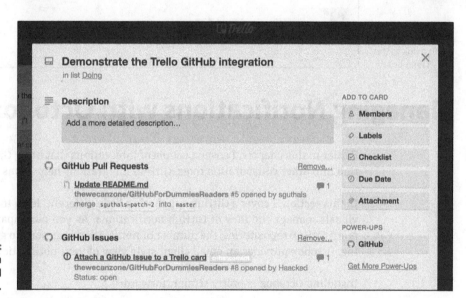

FIGURE 12-11:
Trello card with
an issue and pull
request attached.

When you visit the issue or pull request on GitHub.com, you can see that the attachment is bidirectional. The GitHub issue now has a link to the Trello board, as shown in Figure 12-13.

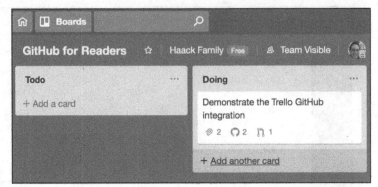

FIGURE 12-12:
Front of a card with an issue and pull request attached.

FIGURE 12-13:
GitHub issue with a link to the Trello board.

Managing Notifications with Octobox

Earlier in this chapter, I cover a couple of integrations that bring GitHub information into other collaboration tools. GitHub integrations help teams work together.

In this section, I cover a GitHub app that's a little different. It's a tool to help individuals manage the flow of GitHub notifications. As you participate in more and more GitHub repositories, the number of notifications can start to get overwhelming. Octobox provides an email client style view of your notifications.

Installing Octobox is pretty straightforward:

1. **Go to** https://octobox.io **and scroll down to the button labeled Install the GitHub App.**

   Some pricing options appear, as shown in Figure 12-14. Octobox is free for open source projects.

2. **Click Install the GitHub App to continue with the installation process.**

3. **Authorize the application the same way you authorized Slack and Trello, earlier in this chapter.**

FIGURE 12-14:
GitHub app
download button
and pricing
options for
Octobox.

After the installation and authorization steps are complete, you're taken to your Octobox inbox. The first time it runs, it takes a moment to synchronize your notifications. When it's done, you should see something like Figure 12-15.

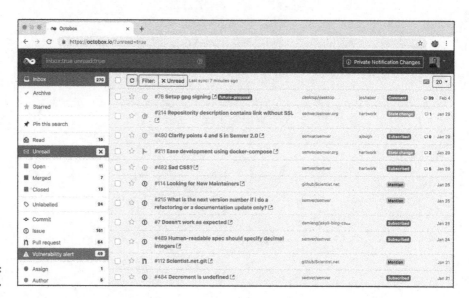

FIGURE 12-15:
Octobox inbox.

After Octobox is installed and synchronized, you can use it to manage your notifications. It allows you to search and filter your notifications by repository, organization, type, action, status, and so on. You can set Octobox to automatically synchronize on an interval in its Settings page. As the status for issues and pull requests change on GitHub, synchronizing Octobox displays those changes in Octobox. Octobox also provides archiving and muting for notifications, which is a nice way of staying on top of notifications, especially if you work on multiple active projects on GitHub.

Chapter **13**

GitHub Workflow Integrations

I n Chapter 12, I show you some ways that you can get information about GitHub repos in other applications. Using applications like Slack and Trello, integrated with GitHub, can especially help you with project management. In this chapter, I show you some ways you can integrate GitHub into your existing development workflow.

A lot of the integrations that I show you in this chapter are open source, which means you can track the development of new features, easily report bugs through GitHub, or even contribute to the project. It also means that each is rapidly changing, so some details may be outdated by the time you read it. What is important is that you know how to find updates and navigate each integration.

Using GitHub for Visual Studio Code

One of the fastest growing editors is Visual Studio Code (VS Code for short). With more than 2.6 million users in the first 12 months in 2017 and over 14 million users today, you have probably either used it or at least heard about it. VS Code is open source, which means you can see the development of new editor features at https://github.com/microsoft/vscode. In 2018, Microsoft and GitHub teamed up to build an open source GitHub editor extension that provides an in-editor pull request experience.

After you install VS Code, you can install this extension by going to the extension marketplace, searching for *GitHub,* and clicking the blue Install button for the GitHub Pull Requests and Issues extension, as shown in Figure 13-1.

FIGURE 13-1:
GitHub extension
for VS Code in the
extension
marketplace.

After you install the extension, you can sign in to GitHub by clicking the GitHub extension icon on the side panel and then clicking the Sign In button, as shown in Figure 13-2.

The sign-in process opens a web browser where you can authorize VS Code to access your GitHub repositories. Click through the prompts, and it redirects you back to VS Code. After you're back in VS Code, you get a pop-up notification asking whether you trust the URL; click Yes.

Interacting with pull requests in VS Code

After you're signed in, when you go to the Source Control tab on the left side of VS Code, you should see all the pull requests associated with this repo. Pull requests are grouped into five different sections:

>> **Local Pull Request Branches:** Ones that you currently have checked out on your machine

>> **Waiting for My Review:** Ones where you are marked as a reviewer

>> **Assigned to Me:** Ones that you're assigned to

>> **Created by Me:** Ones that you created

>> **All Open:** A list of all pull requests that are open for the repository

FIGURE 13-2:
Initiate the
GitHub sign-in
process.

When you unroll a specific pull request, you see the description of the pull request, along with all the modified, added, or deleted files.

Clicking the description of the pull request displays the description as you would see it on GitHub, as shown in Figure 13-3. From this page, you can check out or refresh the pull request; leave a comment; modify reviewers, assignees, labels, or milestones; merge or update the pull request; or complete a review. You have the entire pull request experience right inside of VS Code!

Another feature of this extension is the ability to add inline comments to the diff. Clicking a specific modified file shows you the side-by-side diff, just as it would look on GitHub.com. From here, you can add a comment to any of the modified lines, as shown in Figure 13-4. All these actions are reflected on GitHub.com.

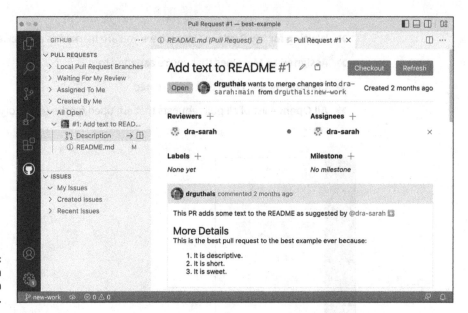

FIGURE 13-3:
Interacting with pull requests in VS Code.

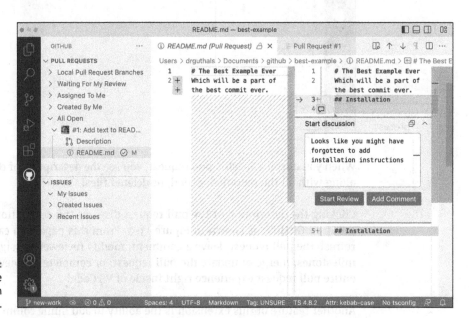

FIGURE 13-4:
Adding an inline comment in VS Code.

Following the GitHub for VS Code pull requests extension

Because this extension is also open source, you can follow the development, report issues, or even contribute to it on GitHub.com. Go to https://github.com/

`Microsoft/vscode-pull-request-github` to find the latest features and documentation for how to use this extension to improve your development workflow.

Using GitHub for Visual Studio

Visual Studio is different from Visual Studio Code. Visual Studio is an integrated deveopment environment (IDE) and is a full-featured application to support developers in writing code, while VS Code is an editor that has an extensive list of extensions that a developer can add to create the experience they need. Visual Studio 2015, 2017, and 2019 have a specific GitHub integration, while Visual Studio 2022 has Git and GitHub integrated out of the box. It is recommend that you install and use Visual Studio 2022; however, if you have to use an older version, this section shows you how to integrate GitHub with Visual Studio 2017.

TIP

If you do install Visual Studio 2022, you can use GitHub very similarly to how to use it in VS Code. Head to `https://visualstudio.microsoft.com/vs/github/` for guided instructions on the newest version control features of Visual Studio 2022.

After you install Visual Studio, choose Tool ⇨ Extensions and Updates. A pop-up window appears with all the extensions you currently have installed, plus the marketplace of additional extensions. If you click Online, the top choice is the GitHub Extension for Visual Studio extension, as shown in Figure 13-5.

Click the Download button and close Visual Studio. When Visual Studio closes, the VSIX Installer starts and ask whether it can modify Visual Studio. Click Yes and Modify, and the extension begins to install. After it installs, click the Connect link in the Team Explorer tab and connect to your GitHub account, as shown in Figure 13-6. When you click the Connect link, a pop-up window asks you to sign in to GitHub. If you have two-factor authentication set up, you're also asked for your 2FA code.

Once connected, you can clone a repo, create a new repo, or sign out from GitHub, all from the Connect page on the Team Explorer pane.

Viewing, creating, and reviewing pull requests in Visual Studio

When you have the project open in Visual Studio that is connected to a GitHub repo, you see additional project options on the home page of the Team Explorer pane, as shown in Figure 13-7.

FIGURE 13-5:
The GitHub for
Visual Studio
extension in the
Visual Studio
Marketplace.

FIGURE 13-6:
The Team
Explorer pane
with the GitHub
Connect section.

Clicking Pull Requests opens the GitHub pane where all the pull requests on this repo are listed, as shown in Figure 13-8. At the top of the list, you can choose to see all the open pull requests, closed pull requests, or just all pull requests. You can also sort by author.

FIGURE 13-7:
The Team
Explorer pane
with additional
GitHub project
options.

FIGURE 13-8:
A list of pull
requests on the
GitHub pane.

If you double-click one of the pull requests, the details open. From here, you can see the description, the target and base branch, the current state, the list of reviews, and the list of files changed. You can also click the View on GitHub link to open a browser window to view this pull request on github.com. Click the Checkout *<branch-name>* button to check out the branch associated with this pull request. Click the Add Your Review link to add your own review with inline and overall comments. When you add a review, you can mark it as comment only, approve it, or request changes. If you double-click one of the changed files, the diff opens in the editor area. If you hover over one of the changed lines, you can add an inline comment, very similar to how it is done in VS Code (refer to Figure 13-4). You can see all of this in Figure 13-9.

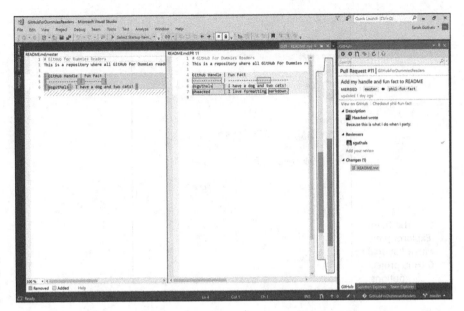

FIGURE 13-9:
Interacting with a
pull request in
Visual Studio.

Following the GitHub for Visual Studio extension

The GitHub for Visual Studio extension is open source, so while it is highly recommended that you install and use the integrated Git experience for Visual Studio 2022, if you want to make improvements on the 2015, 2017, 2019 GitHub integration, you still can! Go to https://github.com/github/visualstudio to find documentation for how to use the GitHub for Visual Studio extension to improve your development workflow.

Using GitHub for XCode

Apple has developed an integration for GitHub in XCode. After you install XCode, you can sign in with GitHub by clicking the Clone from Existing Git Repository button, clicking the arrow next to the message to sign in to an account, and then opening the Accounts menu. Click the + button and choose GitHub, and click Continue, as shown in Figure 13-10. Follow the prompts to sign in to your GitHub account. If you have two-factor authentication set up, you are asked for your 2FA code.

TIP

You need to have a personal access token to sign in to GitHub. You can create one at https://github.com/settings/tokens.

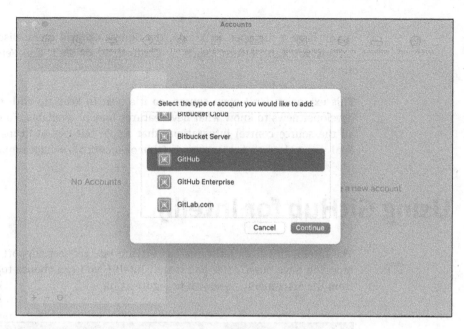

FIGURE 13-10:
Signing in to
GitHub inside
of XCode.

Now when you choose Source Control ⇨ Clone, your GitHub repos load below the box where you can insert a URL to clone. When you click one, you get information, such as the primary language used, the number of forks and starts for this repo, and a link to the README file, as shown in Figure 13-11.

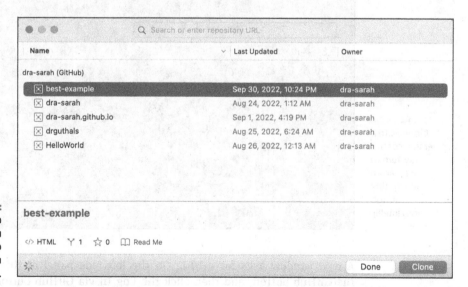

FIGURE 13-11:
List of GitHub
repositories you
have access to
clone from
inside XCode.

If you have a file open in XCode, the Source Control menu displays additional actions you can perform, such as Commit, Push, Pull, and Fetch and Refresh Status.

This extension isn't open source, so it's best to keep up with the latest Apple developer news to know what new features may be available. You can read about all the source control integrations that Apple releases at `https://developer.apple.com/documentation/xcode/source-control-management`.

Using GitHub for IntelliJ

The IntelliJ IDE from JetBrains has GitHub pull request support for any GitHub repo you have open. After you install IntelliJ, you can choose to clone a project from the Start menu, as shown in Figure 13-12.

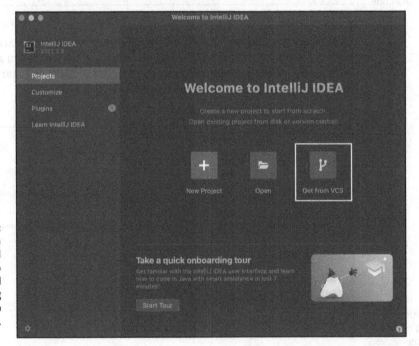

FIGURE 13-12: Clone from a version control system to get started integrating GitHub into IntelliJ.

A new window asks for a URL. On the left side panel of this window, click the GitHub button, and then click the Log In via GitHub button, as shown in Figure 13-13. You're asked to authorize GitHub in the browser, similar to signing in with Visual Studio Code. After you have successfully logged in, all the GitHub repositories that you have access to appear in the clone list, as shown in Figure 13-14.

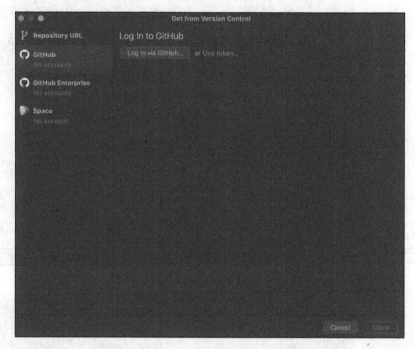

FIGURE 13-13:
The GitHub
log-in window
inside IntelliJ.

After you choose a repository and click Clone, an IntelliJ project window opens with your project. When the project is open in IntelliJ, you can open the GitHub pull request preview by clicking the Pull Requests button on the left pane. A new section opens in the IntelliJ window with a list of the open pull requests, as shown in Figure 13-15. If you click one, the description and list of changed files opens. If you double-click one of the changed files, a diff of that file opens in a new window (see Figure 13-16).

Lastly, you can create a pull request from inside of IntelliJ as well. If you're on a new branch and have already made some changes and committed them to your branch, you can click the + above the list of pull requests, from the same Pull Request button on the left side of your project window shown in Figure 13-15. In the panel, shown in Figure 13-17, specify the title and description of your pull request, and add reviewers, assignees, or labels if you want to.

This GitHub pull request feature is embedded in the IntelliJ IDE, so it's best to follow the IntelliJ blog and documentation for up-to-date information on its development.

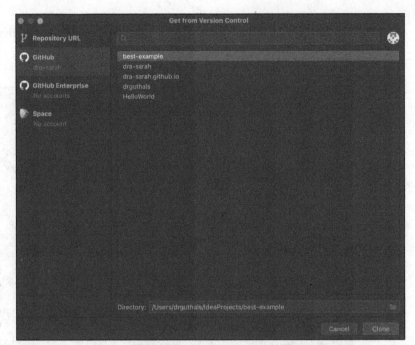

FIGURE 13-14:
List of GitHub repositories you have access to clone from inside IntelliJ.

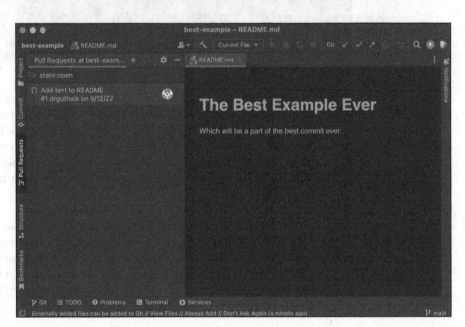

FIGURE 13-15:
A list of open pull requests from inside of IntelliJ.

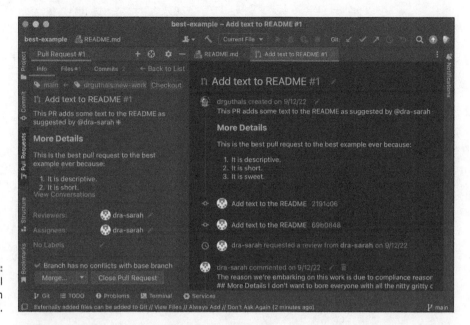

FIGURE 13-16:
Viewing a pull
request from
inside of IntelliJ.

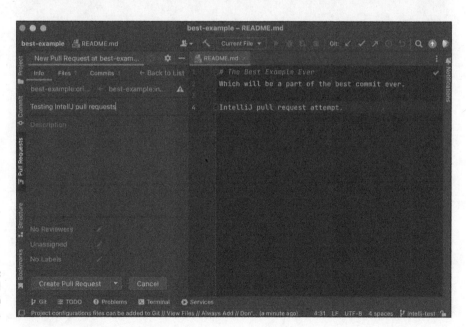

FIGURE 13-17:
Creating a pull
request from
inside of IntelliJ.

Chapter **14**

Personalizing GitHub

D evelopers have a lot of opinions about how they work and invest a lot of time personalizing their tools to work just the way they want. GitHub offers an extensive API that lets developers write tools to interact with GitHub data in a myriad of ways. Also, because GitHub.com runs in a browser, anyone can build browser extensions to customize the experience of using GitHub.

In this chapter, I look at some of the available ways to personalize your use of GitHub.

Using Browser Extensions

Browser extensions can completely customize the experience of using a browser. You can find extensions for every possible scenario you can think of.

In this section, I look at a few useful extensions that work with GitHub. Some of these extensions are available for multiple browsers, but I focus on Google Chrome extensions for brevity.

For a more comprehensive list of browser extensions that work with GitHub, check out this list of awesome browser extensions for GitHub repository made by Stefan Buck at https://github.com/stefanbuck/awesome-browser-extensions-for-github.

Refining GitHub

The Refined GitHub extension is an open source extension that simplifies the GitHub interface and adds some useful features to GitHub.com. The extension is available for the Chrome, Firefox, and Opera browsers.

You can find the source code at `https://github.com/sindresorhus/refined-github`. The README has a link to install the extension. Note that when you install the extension, you grant it the ability to read and change your data on `api.github.com`, `gist.github.com`, and `github.com`. It can also modify data you copy and paste.

After you install the extension, you should see an Octocat icon to the right of the address bar. This icon provides some light customization options. One option lets you define some custom CSS specific to GitHub.com. In Figure 14-1, you can see I added some custom CSS to make the repository name larger and dark red. Note that after you change the CSS, you have to refresh the page to see your changes in effect.

FIGURE 14-1:
Configuring the
Refined GitHub
extension.

Changing the CSS is just a parlor trick compared to the many other enhancements Refined GitHub brings with it.

WARNING

Browser extensions work by manipulating the generated HTML of a website. It often looks for known *landmarks* in the pages (such as an HTML element with a known ID) and then adds its own UI elements, removes elements, or changes elements. However, it does this all outside of the code running on the server that actually generates the HTML. What this means is that if a website such as GitHub.com changes its HTML markup, an extension feature could stop working temporarily until the authors update their extension to adapt to the new change. So if any of these features stop working, try again later after the extension is updated.

The following is a small sampling of enhancements that are on as a default when you have this extension installed:

> **Mark issues and pull requests as unread.** This enhancement adds a Mark as Unread button to the Notifications section of an issue or pull request. Click it to mark the issue or pull request unread, which puts it back in your notifications list.

> **Stop the page jumps from recently pushed branches.** Normally, when you push a new branch, the home page of a repository displays a list of recently pushed branches in a yellow bar above the list of code files. The sudden appearance of this list can cause the rest of the page to jump down to make room for the bar. Refined GitHub displays the list of recently pushed branches in the top right overlaying the location where the Fork button may be. By displaying the branches in an overlay, Refined GitHub ensures that the page doesn't jump down.

> **Adds option to wait for successful checks.** If you have continuous integration (CI) set up for a repository, it can take a while after someone pushes a pull request before all the checks are completed. The CI might be running static analysis or a linter, unit tests, integration tests, and so on. It can be annoying to wait for all those processes to complete after you've reviewed some code and are ready to merge the pull request. Refined GitHub adds a check box that lets you indicate that it should go ahead and merge the pull request after the checks are complete and successful.

> **Reaction avatars show who reacted to a comment.** Typically a reaction comment shows only the reaction and the count for that reaction. With Refined GitHub, you can see who all gave a specific reaction.

Refined GitHub contains many more enhancements big and small. The list here is just the tip of the iceberg. In Figure 14-1, notice that there is also a place to disable features that are a part of the extension, as well as a place to put your GitHub personal access token so that the extension can work on private repos as well.

Taking a GitHub selfie

As an open source project maintainer, I am ecstatic when someone comes along and submits a pull request for a project. I'm happy when someone opens an issue that identifies a problem I didn't know about. I feel gratitude for the folks who take time out of their day to help out my project.

And sure, I could use words to communicate my gratitude, but they always seem to fall short of my true feelings. If the old saying that a picture is worth a thousand words is true, how many words is an animated gif worth?

GitHub Selfie is an open source browser extension (https://github.com/thieman/github-selfies) that adds a Selfie button to the comment field that lets you take a selfie using your computer's camera. You can choose to take a still picture, but where's the fun in that? It also provides an option to take an animated gif.

Figure 14-2 shows a ridiculous self-portrait.

Adding a selfie to express your gratitude is a small detail, but adds a nice warm personal touch when you're working with people from all over the world.

GitHub Apps and Probot

Chapter 12 covers a few integrations that connect GitHub to other applications, such as Slack and Trello. What those integrations have in common is they were implemented as GitHub apps.

Apps on GitHub let you extend GitHub in powerful ways. GitHub apps are web applications that can respond to events on GitHub. These event subscriptions are called *web hooks*. When an event occurs on GitHub that the app is interested in,

GitHub makes an HTTP request to the app with information about the event. The app can then respond to that event in some manner, often resulting in a call back to GitHub via the GitHub API.

In this section, I walk you through building a simple GitHub app that brings a bit of levity to your issue discussions. There's an old meme in the form of an animated gif with a little girl who asks the question, "Why don't we have both?" The typical application of this meme is in response to a question that presents a false dichotomy. In other words, when someone presents a question with two choices, someone might respond with this image.

In this section, you create a GitHub application that will automatically do this as a fun exercise.

Introducing Probot

GitHub apps are web applications that need to listen to HTTP requests. You have a lot of important details to get just right when building an HTTP request, such as what is the format of the data posted to the app? All these details can be confusing and time consuming to get correct when building a GitHub app from scratch. Knowing where to start is difficult.

GitHub's Probot framework comes in handy when getting started with a GitHub app. Probot handles much of the boilerplate and nitpicky details of building a GitHub app. It's a framework for building GitHub apps using Node.js. It provides many convenience methods for listening to GitHub events and for calling into the GitHub API.

Probot makes it easy to build a GitHub app, but it doesn't solve the problem of where to host the app.

Hosting the app

A GitHub app can take many forms. It could be a Node.js app running in Heroku, a serverless function in GitHub Actions, or any other cloud provider — it doesn't matter. It just needs to be persistent and available via the public Internet so that GitHub can reach it with event payloads.

Setting all that up can be time consuming, so for my purposes, I developed the simple Probot app locally using the command line interface (CLI) and then used Glitch to deploy.

Introducing Glitch

Glitch (https://glitch.com) is a hosting platform for web applications that removes a lot of the friction with getting a web app up and running. Any app you create in Glitch is live on the web from the beginning. You don't have to think about how you plan to deploy the code because any change you make is auto-saved and automatically deployed.

Glitch focuses on the community aspect of building apps. Every file can be edited by multiple people in real-time, in the same way you might edit a document in Google Docs. And every project can be remixed by clicking a button. This encourages a lot of sharing of code and learning from each other, which comes in handy when we build our own GitHub app.

Before you continue, make sure to create an account on Glitch if you don't have one already.

Creating a Probot app

Follow these steps to get started:

1. **Create a local Probot app by opening a command line and typing:**

```
npx create-probot-app why-not-both
```

2. **Type y to create the Probot app locally, and then answer the questions in the command line.**

For example:

```
? App name: why-not-both
? Description of app: A probot app to reply to issues
? Author's full name: drguthals
? Which template would you like to use? basic-js => Comment
  on new issues
```

3. **Once the files are created, open the folder in Visual Studio Code (or your editor of choice) and edit the index.js file with the following code:**

```
/**
 * This is the main entrypoint to your Probot app
 * @param {import('probot').Probot} app
 */
module.exports = (app) => {
  // Your code here
  app.log.info("Yay, the app was loaded!");
```

```
app.on("issues.opened", async (context) => {
  const message = context.payload.issue.body;
  if (message.indexOf(' or ') > -1) {
    const issueComment = context.issue({
      body: "![The why not both girl](https://media3.
giphy.com/media/3o85xIO3317RlmLR4I/giphy.gif)",
    });
    return context.octokit.issues.
createComment(issueComment);
  }
});


// For more information on building apps:
// https://probot.github.io/docs/


// To get your app running against GitHub, see:
// https://probot.github.io/docs/development/
};
```

This code listens to new issue comments, looks for the word *or* surrounded by spaces, and if it finds it, creates a new comment with a markdown image.

TIP

This approach is not very smart. You can find a slightly better approach at https://git.io/fhHST. It would be even better if I could employ artificial intelligence (AI) in the form of natural language processing (NLP). But that's beyond my skillset and out of the scope for this book.

4. **Save the file, and back in your terminal, start the server with the following command:**

```
npm run start
```

5. **Once the server starts, open a browser and point it to localhost:3000.**

   The Register GitHub App button is now part of the interface, as shown in Figure 14-3.

6. **Click the button and follow the instructions to register the app to your GitHub account and whatever repositories you want to associate it with.**

7. **After registering, shut down the local server by pressing ⌘-C or Control+C.**

8. **Open the .env file and notice that GitHub set critical environment variables:**

   ● WEBHOOK_PROXY_URL

   ● APP_ID

FIGURE 14-3:
The Probot app
with a button to
register the app
with GitHub.

- PRIVATE_KEY
- WEBHOOK_SECRET
- GITHUB_CLIENT_ID
- GITHUB_CLIENT_SECRET

9. **Test your Probot app by restarting the npm server with the following command:**

```
npm run start
```

10. **Go to the one of the repositories that you authorized and create a new comment with the word _or_ in it.**

Watch the Probot app automatically responds with a gif of the "Why not both" meme, as shown in Figure 14-4.

Make sure you shut down your npm server by pressing ⌘-C or Control+C in your terminal.

Pushing the Probot app to GitHub

Now that you have verified the Probot app works, it's time to get it hosted in the cloud. To get started, you need to push your code to GitHub. In your Probot root folder, initialize a Git repository with the following command:

```
git init -b main
```

Next, stage your initial commits with the following command:

```
git add . && git commit -m "initial commit"
```

Finally, push your repository to GitHub. To do this, type the following command into your terminal:

```
gh repo create
```

This kicks off the GitHub repository creation flow:

1. Using your arrow keys, choose Push an Existing Local Repository to GitHub.

2. Press Enter to choose . as the location for the repository (this means the current folder).

3. Press Enter to choose the name of the folder as your repository name, or type a new repository name.

4. Add a short description, such as "A simple Probot app".

5. **Using your arrow keys, choose Public, Private, or Internal for the repository visibility.**

6. **Type** Y **to create a remote to the GitHub repository with your local repository.**

7. **Press Enter to choose origin as the name of the remote.**

8. **Press Enter to choose Yes to commit your local changes to the remote GitHub repository.**

You can now navigate to your repository on GitHub and your code should be there. You can see mine at `https://github.com/drguthals/why-not-both`.

Hosting your Probot app on Glitch

Now that you have confirmed your Probot app works when it is locally hosted and you have published your code to GitHub, you can easily use Glitch to host your Probot app. Go to `https://glitch.com`, sign in, and choose New Project ⇨ Import from GitHub.

Type your repository name to the text field, making sure to include `.git` at the end. For example, for my Probot app I would type `https://github.com/drguthals/why-not-both.git`.

Once Glitch loads your code in the online Glitch editor, you should see all the files from your repo have been imported into Glitch and the `README.md` file is open in the code editor.

You now have to register this new Probot app with GitHub. To do this, follow these steps:

1. **Click Preview, found at the bottom of the Glitch editor.**

2. **Choose Preview in a New Window.**

3. **Click the Register GitHub App button.**

4. **Name your GitHub app.**

5. **Choose which organizations, accounts, and repositories to install your app on.**

6. **Click Install.**

Head to your repository and make sure the Glitch-hosted Probot app works by opening a new issue. You should get a response, as shown in Figure 14-5.

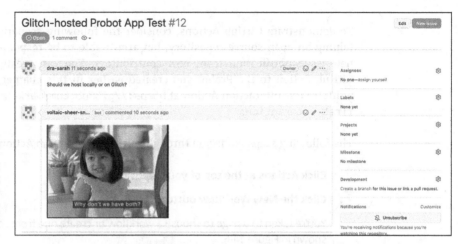

FIGURE 14-5:
The same Probot
app in action on a
GitHub issue,
but this one is
hosted on Glitch.

Taking Action with GitHub Actions

In the previous section, I walk through personalizing GitHub by creating a GitHub app. This required that you host your app outside of GitHub. It turns out GitHub has a feature that removes the need to host your app outside of GitHub, which can reduce the number of moving parts when extending GitHub. This feature is called GitHub Actions.

GitHub Actions makes it possible to create custom workflows on GitHub. It lets you implement custom logic to respond to events on GitHub. In the previous section, you wrote a GitHub app to do that. With GitHub Actions, you don't need to build a custom app. You can build workflows using existing actions that others have written, or you can write your own actions that run in a Docker container.

EXPLORING GitHub ACTIONS

There are pre-written GitHub Actions — some really useful and complex — for almost every part of your developer workflow on GitHub. You can find a list of them at https://github.com/marketplace?type=actions. You can also follow some tech folks who write and open source more specific GitHub Actions.

You can also write custom Actions. How to do that is beyond the scope of this book, but you can look at the Actions docs at https://docs.github.com/actions.

To demonstrate GitHub Actions, consider the following scenario. You're maintaining an open source repository, but aren't able to be respond to every single new issue notification from new contributors. You can create a very similar GitHub Action to the Probot you created earlier in this chapter using the pre-written First Interaction Action at `https://github.com/marketplace/actions/first-interaction`.

The following steps guide you through setting up a GitHub Action:

1. **Click Actions at the top of your repository.**

2. **Click the New Workflow button.**

   You're taken to a page to choose a workflow or create one from scratch, as shown in Figure 14-6.

3. **Search for the Greetings Action and click Configure.**

   This creates a `greetings.yml` file in your `.github/workflow` folder, as shown in Figure 14-7.

4. **Click Start Commit and commit the new `greetings.yml` file to the `main` branch.**

5. **Invite a friend to open a new issue on your repository for the first time.**

   Right when the issue is created, if you head back to the Actions tab you can see the workflow running, as shown in Figure 14-8. When it completes, a comment is added to the new issue, as shown in Figure 14-9.

FIGURE 14-6: Options for choosing a workflow.

FIGURE 14-7:
Create a
greetings.yml file
to install the
GitHub Action.

FIGURE 14-8:
The GitHub
Action running.

FIGURE 14-9:
The comment
added to a new
issue by the
GitHub Action.

6

The GitHub Ecosystem

Chapter **15**

Exploring the GitHub Marketplace

I n the three chapters of Part 6, I look at multiple different ways of extending GitHub and customizing the GitHub experience. Many tools extend or integrate with GitHub. A good way to find tools to use with GitHub is the GitHub Marketplace.

Introducing the GitHub Marketplace

The GitHub Marketplace (https://github.com/marketplace) is a directory of tools and apps grouped in the following categories:

>> API (application programming interface) management

>> Chat

>> Code quality

>> Code review

>> Continuous integration

>> Dependency management

- » Deployment

- » IDEs (integrated development environments)

- » Learning

- » Localization

- » Mobile

- » Monitoring

- » Project management

- » Publishing

- » Recently added

- » Security

- » Support

- » Testing

- » Utilities

The Marketplace is a great way to find an app for every situation on GitHub. Purchasing or installing apps through the Marketplace has two key benefits: ease of billing and installation and the vetting process.

Billing made easy

For apps in the GitHub Marketplace that require payment, installing the app through the Marketplace is a streamlined flow because you can use your GitHub payment info. That way, you're not dealing with five different payment providers when purchasing five different apps to use with GitHub.

If you have a free GitHub account, you may not have set up your payment information in GitHub. To set up a payment method, click your avatar in the top-right corner of GitHub.com and click Settings. From this page, click Billing and Plans from the list on the left side. Here you can click the Add Payment method, as shown in Figure 15-1.

The Marketplace vetting process

One of the benefits of installing an application from the Marketplace is that these apps must meet certain requirements before GitHub lists them in the Marketplace. The requirements help ensure a higher standard of quality and security with the apps, helping ensure that these apps are useful (no Fart apps) and are secure.

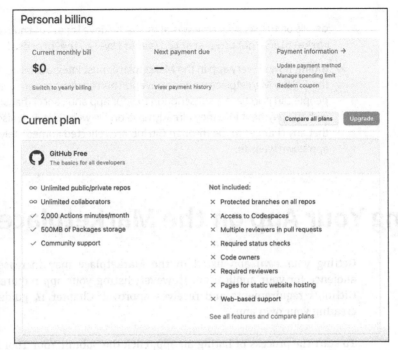

WARNING

At the moment, a GitHub Action doesn't require any review to be listed in the GitHub Marketplace, which means installing an Action from someone you don't know may be a bit riskier.

An app must meet four main categories of requirements before being listed in the Marketplace (`https://developer.github.com/marketplace/getting-started/requirements-for-listing-an-app-on-github-marketplace`):

» **User experience:** This brief set of nine requirements includes things like the app must have a certain number of users and installs already. It also includes some requirements around the behavior of the app, such as the app must include links to documentation, it can't actively persuade users away from GitHub, and it must provide value to customers.

» **Brand and listing:** This set of guidelines and recommendations center around the branding of your app and your app's listing. Every app must include its own logo. If the app makes use of GitHub's logo, it must follow GitHub's Logos and Usages guidelines. The brand and listing section on the Requirements page has links to further logo and description guidelines. As you can see, GitHub takes listing apps in the Marketplace seriously.

» **Security:** GitHub conducts a security review of apps before listing them in the marketplace. A separate document with security best practices and more details on the security review is at https://developer.github.com/marketplace/getting-started/security-review-process.

>> **Billing flows:** Every app in the Marketplace must integrate billing flows using the GitHub Marketplace webhook event. This requirement ensures that people can purchase a subscription to your app and cancel that subscription with the payment info they already have on file with GitHub. It also ensures that any changes made through GitHub are reflected immediately on the app's own website.

Listing Your App on the Marketplace

Getting your own app listed in the Marketplace may increase the potential audience for your application. However, listing your app requires that it meets GitHub's requirements and receives approval. Chapter 14 guides you through creating your own app.

To start the process of listing an app, click the Submit Your Tool for Review link at the bottom of the Marketplace landing page or navigate to https://github.com/marketplace/new in your browser.

This page lists your applications that you can turn into Marketplace listings, as shown in Figure 15-2. Notice the Why Not Both app I created in Chapter 14 is listed here.

FIGURE 15-2: Your applications that you can turn into Marketplace listings.

Click the Create Draft Listing button next to the app you want to list on the Marketplace to start the process. This takes you to a page where you can enter a name for the listing and choose one of the marketplace categories for your app listed earlier in the chapter, as shown in Figure 15-3.

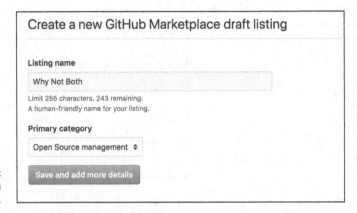

Create a new GitHub Marketplace draft listing

Listing name

Why Not Both

Limit 255 characters. 243 remaining.
A human-friendly name for your listing.

Primary category

Open Source management ⬦

Save and add more details

FIGURE 15-3:
Filling out a form
to list an app.

TIP

If you save the draft of your listing but happen to close your browser, you can get back to your listing by going to https://github.com/marketplace/manage in your browser.

After you fill in these details, click the Save and Add More Details button to save a draft of your listing and move on to the next set of steps, as shown in Figure 15-4.

These steps include

1. **Add your contact info.**

   This info is a set of three email addresses: Technical lead, marketing lead, and finance lead.

2. **Fill out your listing description.**

   This area is where you fill out more details, such as a product description, logo, and screenshots. The information here will be displayed on the Marketplace page for your application.

3. **Set up plans and pricing.**

   This is where you can set up one or more pricing plans, including the option to create a free plan, a monthly plan, or a monthly per user plan. You can also specify whether a plan includes a 14-day free trial.

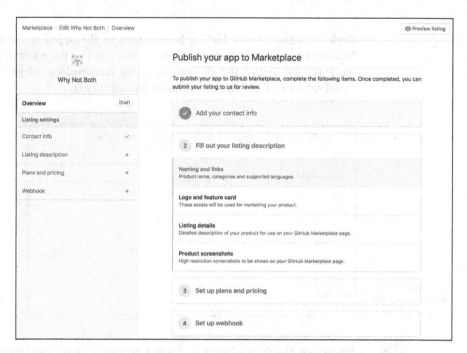

FIGURE 15-4:
Steps to fill out a
Marketplace
submission.

4. **Set up webhook.**

   This step allows you to specify a URL where Marketplace events will be sent via an HTTP POST request. The webhook will send you information about events, such as purchases, cancellations, and changes such as upgrades and downgrades.

5. **Accept the Marketplace Developer Agreement.**

   To list your app in the marketplace, you have to accept the Marketplace Developer Agreement.

6. **Click the Submit for Review button.**

   GitHub employees review your submission to make sure it meets the requirements to be listed in the Marketplace. The result of the review of your submission will be emailed to you.

Considering Common Apps to Install

In the section "Introducing the GitHub Marketplace" at the beginning of this chapter, I list the categories of apps that are available on the Marketplace. In this section, I describe some of the most common and useful apps that you may want to consider installing.

Continuous integration

Continuous integration (CI) apps automatically build and test your code every time you push it to GitHub. If you have a CI app, such as AppVeyor, installed on your repository, you see the status of the check at the bottom of each pull request, as shown in Figure 15-5.

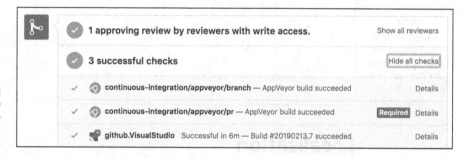

FIGURE 15-5: AppVeyor CI app example on the GitHub for Visual Studio repository.

If you're the owner of the repository, you can also specify whether checks have to pass before the branch can be merged into the main branch. Just head into the Settings tab. If you have any rules on the main branch already, click Edit; otherwise, click Add Rule. From there, you can scroll down and select Require Status Checks to Pass before merging.

Code quality

Code quality apps automatically review your code with style, quality, security, and test-coverages checks. These apps can be really useful for ensuring your code is kept to a high standard. With well-styled and quality code, you're less likely to introduce or miss bugs. For example, if you require that all curly braces are on new lines and indented with one tab per nested brace, you're likely to be able to spot when something is incorrect. For example, Rubocop checks the style of your Ruby code while it's building and provides you with style feedback, such as casing for method names.

Another useful type of code quality apps is the code coverage apps, such as Codecov. Shown in Figure 15-6, Codecov and apps like it comment on pull requests with how much of the code is covered by test scenarios, helping to ensure your code remains well tested.

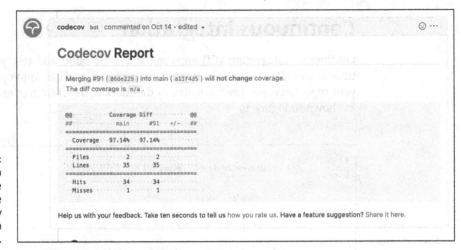

FIGURE 15-6:
Codecov app
example on the
GitHub package
for the codecov
example-python
repository.

Localization

Localization apps can make publishing your app in many languages easier. For example, the Crowdin app links your repository to a Crowdin account where people from around the world can help you translate your documentation and any written words in your software (for example, on buttons or in menus). With more than 20,000 people contributing to translations, the Crowdin app automatically opens a pull request on your repository with new translations when it's reached a threshold of accuracy, still giving you a chance to review and merge. For open source projects, Crowdin is free!

Monitoring

Monitoring apps help measure performance, track errors, and track dependencies in your code. For example, Greenkeeper is a real-time notification app that gives you updates and changes for JavaScript dependencies. Figure 15-7 shows Greenkeeper in action, opening a pull request to update eslint to the latest version.

Dependency management

Modern app development today is heavily dependent on public package managers for pulling in and managing dependencies. A typical app may have dozens, if not hundreds, of dependencies. Tracking which of these dependencies are up-to-date can be difficult. GitHub apps such as Dependabot check to make sure your dependencies are up-to-date and submit pull requests to update the ones that are not.

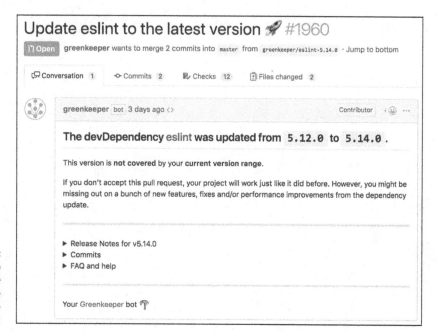

FIGURE 15-7:
Greenkeeper app
example on the
GitHub package
for VS Code
repository.

Sometimes you don't want all your dependencies on a public package registry. For example, if you work in an enterprise, you may have internal packages that should remain private. A private package registry tool, such as MyGet, is useful in this case. MyGet works with NuGet packages and lets you set up a policy where pushes to a particular branch initiate a build and deploys the branch to a custom NuGet feed hosted on MyGet.

Testing

Testing software is an important part of the software development lifecycle. Good testers develop test plans to ensure that testers do a good job of testing each release. Managing test plans and keeping track of the status of test runs is an important part of quality assurance. The TestQuality app integrates with GitHub to helping developers and testers create, run, coordinate, and monitor software testing tasks.

Learning

TIP

A great way to learn GitHub is to install the GitHub Learning Lab from the Marketplace. Learning Lab installs a bot that walks you through interactive lessons on how to use GitHub through a set of tasks that you complete. The lab is free and lets you take as many courses as you like at your own pace.

Chapter **16**

GitHub and You

itHub is often described as a social network for developers. Throughout this book, I show you that GitHub is much more than a social network. It's an essential set of tools for working on code together. Even so, the social network aspect is still an important part of GitHub. It may well be a key reason for its success. When GitHub was created, there were existing source control hosts. Pretty much all these hosts were focused around projects. GitHub turned this project-focus approach on its head and made people the focus. You can see it in the URL structure where every repository has the name of the user or organization before the repository name.

In this chapter, I dig into the social network aspect of GitHub. I look at how you represent yourself in this network and how you can get involved in the online community.

Understanding Your GitHub Profile

Every user on GitHub has a profile page. Figure 16-1 shows my profile at https://github.com/drguthals.

Your profile page represents you on GitHub. When you open an issue or submit a pull request to a new repository, the maintainers are likely to take a look at your profile to get a sense of you.

REMEMBER

You can create a profile description by creating a repository with the same name as your username, as described in Chapter 2.

Not only that, your GitHub profile can serve as a portfolio of your development work. It provides some insight into your interests, experience, and ability as a software developer, as shown in Figure 16-2. Many companies who are hiring will take a look at your GitHub profile.

WARNING

Many companies give heavy weight to applicants who have a GitHub profile. However, this biases against people who don't have the benefit of working at a company that uses GitHub. It also biases against those who don't have free time to work on open source. GitHub, Inc. itself doesn't require a strong GitHub profile to apply for a job. More and more companies are taking an approach where they'll look at your GitHub profile if you have one, but won't hold it against you if you don't.

Profile picture

The first thing people visiting your profile will notice is your profile picture. This pic is associated with all your activity on GitHub. When you create an issue or a pull request or leave a comment, your profile pic is right there next to it.

TIP

Your photo is an important part of your GitHub identity so make it reflect your personality, whether it's a picture of you smiling warmly, a photo of a cartoon character, or a landscape photo.

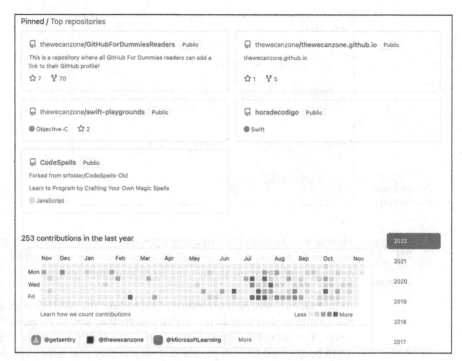

Status message

To the right of your profile pic is a status message you can use to communicate something about yourself to the entire community. In Figure 16-1 you can see that my status was set to an emoji of a smiley with a bowtie. When you hover over the status emoji, the status message pops up if there is one. For some, it's an outlet to say something funny or meaningful. But for others, it's used for practical purposes. For example, if you're a maintainer of a popular project, you may want to let the world know that you're busy if you plan to be away from GitHub for a while. That sets the expectation that you may be slow to respond to new issues. Click your status message to bring up the option to change it. Figure 16-3 shows the status message dialog box with a busy message specified.

Personal info and bio

GitHub displays your bio and information that you choose to show the world under the status message. Click the Edit button to change your bio, company name, location, and URL. This area is a good opportunity to tell the world a bit more about yourself and link to your personal blog or website.

REMEMBER

All this information is completely optional. By entering it, you give GitHub permission to display this information wherever your user profile appears. You can delete the information at any time by editing your profile.

In this section, you also sometimes find badges that GitHub adds to users' profile. For example, if you visit the drguthals profile, you can see that I have the Pro badge. I'm definitely a pro GitHub user, but that's not what that badge means. A Pro badge on a GitHub profile just means that that user pays for the individual Pro subscription. (You can see different pricing models at https://github.com/pricing.)

Pinned repositories

By default, GitHub shows a selection of your most popular repositories on your profile page, but those repositories may not represent what's important to you. Click Customize Your Pinned Repositories to select up to six repositories to pin to your profile page.

Pinned repositories can be useful from three perspectives:

>> **You:** Pinned repositories are useful to you because they're direct links to the repositories you care about most. Pinning repositories can make it quick and easy for you to get there, instead of having to create bookmarks for each of them.

>> **Other GitHub users:** When other people visit your profile on GitHub, the first introduction to the kind of work you do is the set of your pinned repositories. This gives folks a sense of what you're interested in. It also highlights the areas where you have experience. Listing areas of experience can be important if you're just starting to contribute to a new open source project because maintainers will often visit your profile page to get a sense of who you are.

> >> **Companies:** Hiring managers and recruiters may look at your profile when you apply to a job to get a sense of your experience. Pinning some projects that you have most contributed to or are most proud of to the top of your profile can help them to get an accurate picture of you.

REMEMBER

You can also always change up your pinned repositories based on what you're doing. For example, if you're applying for a new position, you may want to refresh what is pinned to be more specific to the role you're applying to. And if you're trying to learn something new, you may want to pin repositories that you're currently engaged with. Think of your entire profile page as a living document — one that should be updated as your goals and interests change.

Contribution graph

The *contribution graph* is a grid of squares 7 squares high and 52 squares in length representing each day of the past year. Each square is filled in with a color that represents your contribution level on that day. If you didn't make any contributions, the square remains gray. If you made some contributions, the square ranges from light green to dark green, depending on how many contributions you made.

Issues and pull requests count as contributions on standalone repositories. Commits to a standalone repository's default branch (typically `main`) or to its GitHub pages branch (typically `gh-pages`) count toward your contribution graph.

TIP

If you fork a repository, an Issues tab won't appear at the top of the repo because, typically, the goal of a forked repository is to contribute back to the original open source project. You fork the repository, make your changes, and open a pull request that targets the original repository. If you really want to have issues in a forked repository, you can turn them on in the Settings tab, though I don't recommend it as you should be tracking your progress on an issue in the original repository. Because the intended behavior of a fork is to contribute back to the original repository, issues on the forked repository and pull requests that do not target the original repository do not count toward your contribution graph. The thought here is that you haven't yet contributed to the main project; the fork is kind of just like your own private branch and the true contribution is made after you've added thoughts to an issue or merged code in on the original repository.

The contribution graph is one of GitHub's more controversial features. Two common concerns are raised. The first is that it promotes unhealthy behavior in that many people attempt to keep long streaks of activity going. Many people take pride in having activity on every single day of their graph, even weekends. Though you may be working on something that makes you happy on the weekends, which is okay, it's also very important to recognize that it's not — and shouldn't be — expected that you are coding every single day of your life. Some of the most important aha! moments have come from taking a break and gaining a new perspective.

The other concern is that other people draw bad conclusions about a person's ability as a developer based on the activity graph. For example, someone may look at a developer's contribution graph, see very little activity, and conclude that they're not very productive.

WARNING

Drawing any conclusions from other people's contribution graphs doesn't make sense. A contribution graph should be useful only to yourself as a fun way to see your history of activity. For one thing, the contribution graph is easily gamed. In fact, a repository provides code for doing pixel art using your contribution graph at `https://github.com/nikhilweee/github-activity-art`. It'd be easy to use that tool to make your contribution graph completely dark green. The contribution graph isn't meant to be used as a definitive productivity metric.

Also, contributions to private repositories may not be showing up in a contribution graph. Click Contribution Settings to change that setting. Figure 16-4 shows an example of both enabling the display of private contributions as well as an activity graph. Unlike the contribution graph, which shows how much activity you have, the activity graph shows where your activity occurs.

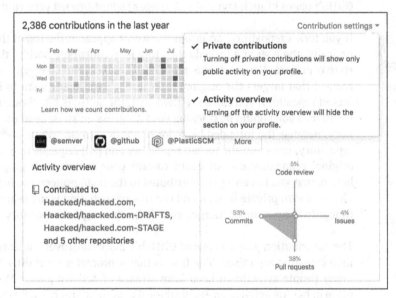

FIGURE 16-4:
Displaying private contributions and the activity graph.

Contribution activity

Underneath the contribution graph is the contribution activity timeline. This timeline of your activity on GitHub goes all the way back to your first commit. It can be nostalgic to go back to the beginning of your activity.

WARNING

In some cases, you may have activity in your timeline that shows up before you joined GitHub. You may even have activity that occurred before GitHub was created! How is that possible? The activity timeline includes Git activity in your repositories based on the timestamp of the Git commits. It's possible to import a repository into GitHub that was created before GitHub existed, which would cause you to have activity prior to GitHub's creation.

Starring Repositories

When you visit a useful repository, you can star it by clicking the star in the top-right corner of the repository page. A star is not only a compliment to the repository owner, but also serves as a bookmark of sorts for the repository.

On your profile page, click the Stars tab to see a list of all the repositories that you've starred. This list is viewable by others who happen upon your profile page. Exploring the repositories others have starred is a great way of discovering interesting new projects.

TIP

If you're curious about which repositories have the most stars on GitHub, use GitHub search (`https://git.io/fhdkx`) to sort users by follower count. This shortened URL shows the top starred repositories on GitHub.

TIP

You may be curious about which of your repositories have the most stars. Right now, you can't list your repositories in order of stars. However, `https://profile-summary-for-github.com/user/haacked` provides a nice visualization of your profile and includes which of your repositories have the most stars (see Figure 16-5). Just replace the username *haacked* with your own to see your profile.

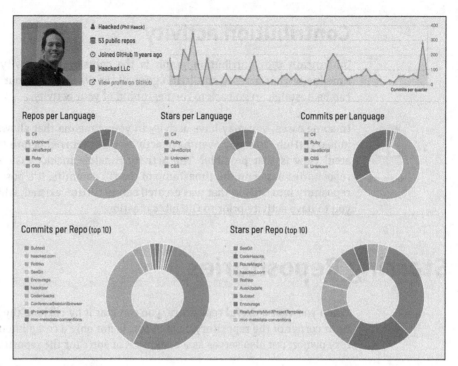

FIGURE 16-5:
Profile
visualization.

Following Users

When you visit the GitHub profile of another user, a big Follow button appears underneath the profile picture. Click the Follow button to subscribe to notifications about the user's activity in your dashboard at `https://github.com`. Who you follow also feeds into GitHub's recommendation system. For example, if someone you follow stars a public repository, that repository may show up in the Discover Repositories section of your dashboard as a recommendation.

You can see all the users you follow by clicking the following link on the left of your profile page. You can see all the people who follow you by clicking the followers link.

TIP

If you're curious about who are the most followed people on GitHub, you can use GitHub search (`https://git.io/fh7P7`) to sort users by follower count. This shortened URL shows the top followed people on GitHub. It may come as no surprise that Linus Torvalds, the creator of Linux and Git, is the most followed person on GitHub.

Chapter **17**

Attending Events

S tarting a career as a software developer is very challenging, especially if you do it on your own. This is one reason why GitHub is such an essential tool for software developers. As the biggest source code host in the world, it's also the biggest software developer community. By participating in GitHub, you become connected to the world's experts and best teachers for any technology you may be interested in.

Most people who use GitHub barely scratch the surface of what GitHub offers. By reading this book, you're a step ahead of many other developers. The knowledge of how to use GitHub will serve you well. But writing code on GitHub only scratches the surface of a rewarding career as a developer.

In this chapter, I look at how to get two steps ahead in your career by encouraging you to step away from the keyboard for a moment and meet with other developers in-person. Attending events is a great way make connections that will benefit you during your entire career. And speaking at events is a great way to grow in your career even more. It may feel daunting, but everyone has something to offer others, even those just starting out.

Exploring Types of Events

Many kinds of events focus on software developers. They range from the informal meet-up or user group to the structured multiday international software conference. In this section, I cover the most common types of events and what to expect at each.

Meet-ups and user groups

A *meet-up* or *user group* is an informal gathering of developers to cover a topic. Many are scheduled monthly and hosted by a local company or software interest group.

These events tend to be a great way to dip your toe into software developer events. They tend to be small gatherings of people in your area. Each month features a local speaker who talks about a topic relevant to the group. (Some user groups and meet-ups bring in speakers from outside on occasion, but typically they focus on highlighting local speakers.)

TIP

Meetup.com is a great way to find a meet-up relevant to your interests. For a list of JavaScript meet-ups in Seattle Washington, go to https://www.meetup.com/topics/javascript/us/wa/seattle. You can search for meet-ups by location on meetups.com.

A few examples of local meet-ups include

- **Brooklyn JS:** http://brooklynjs.com in Brooklyn, New York
- **.NET São Paulo:** www.meetup.com/dotnet-Sao-Paulo in São Paulo, Brazil
- **SD Ruby:** www.sdruby.org in San Diego, California

Regional conferences

A *regional conference* is a relatively small conference where speakers and attendees outside of the local area are welcome, but the focus of the conference is to provide a venue for local developers and speakers to connect and present their work.

Often these conferences are one or two days. Many have a single track of talks, or two at most. They're a step up in size and structure from a meet-up and typically occur once a year, as opposed to monthly.

Some of them often offer workshops either before or after the conference. These workshops usually cost extra, but offer more in-depth training for a specific skill-set or technology. For example, you can often find a full-day workshop dedicated to improving your Git skills. If you can afford it and find one that teaches a skill you want to improve, workshops are often worth the investment.

Some great examples of local conferences include

>> **Caribbean Developers Conference:** https://cdc.dev/ in the Caribbean each year

>> **JSConf Chile:** https://jsconf.cl/ in Santiago, Chile

>> **London Tech Week:** https://londontechweek.com/ in London, UK

Hackathons

A *hackathon* is very different from a conference. While conferences focus more on having speakers teach a topic through a talk, hackathons focus on building. A hackathon is an event that may last several days where groups of people form teams to work together to collaboratively write code to solve some sort of problem.

The usual format is some sort of problem is presented and teams are tasked with building a solution. The technology stack they may use is often dependent on the focus of a hackathon. For example, a mobile development hackathon requires that attendees build a mobile app to solve the problem.

Hackathon is a portmanteau of the words hack and marathon. Many take the marathon aspect to the extreme by having teams work around the clock with very little sleep. Others try to create a balance of working hours and sleeping hours by forcing contestants to leave the workspace.

TIP

Hackathons are often very inclusive of beginners. You don't always have to have a team when you sign up for a hackathon. Often, you can find one when you get there. It's best to check out the FAQ for the specific hackathon to learn more about the details.

One of the largest, worldwide hackathons is targeted to college students. It's the Microsoft Imagine Cup (https://imaginecup.microsoft.com/). Winners of the Imagine Cup can win mentorship from Satya Nadella (Microsoft CEO), travel to the world championship, and receive Azure grants and $100,000.

Attending hackathons can be a great way to be introduced to a new technology. The goal isn't to design and implement a final product, but rather to hack together bits and pieces to make progress on an idea that you have. The end product should look more like a prototype than a polished application. Often times, hackathons have mentors who know a particular technology that you can learn from. Think of a hackathon as a dedicated time and place to experiment and learn.

TIP

Though attending a formal hackathon provides you with mentors, a space, and sometimes prizes, you can also always get together with friends to do one on your own, too! Just pick a time, place, and goal and try to hack together a prototype of an idea you have! It doesn't hurt to give it a shot!

Major conferences

A major conference tends to be large and draw attendees from all over the country, if not the world. Attendee counts tend to be in the thousands. Attending one of these conferences requires a bit more up-front planning. It's not just arranging your flight and hotel. These conferences tend to have many tracks, so for any given time slot, you may have to choose which talk you want to see.

Like a regional conference, major conferences often offer an array of workshop offerings before or after the conference. In addition to workshops, many also include hands on labs during the conference. Labs are usually included in the price of the conference and offer a great chance to actually try out the technologies you're hearing about at the conference.

Many of these conferences are thrown by large technology companies, such as Microsoft's Build conference and Apple's WWDC.

A few examples include

» **All Things Open:** https://allthingsopen.org/ (location changes each year)

» **Build:** https://mybuild.microsoft.com/ in Seattle, Washington

» **WWDC:** https://developer.apple.com/wwdc22/ in San Francisco, California

Knowing What to Expect at Events

Events can vary widely in terms of what to expect, but they all have a few commonalities. The most obvious thing to expect is that other developers will be there. Not everyone has the benefit of living in a tech hub. If you live outside of a tech hub, being a developer can feel solitary. If you work at a company that is not primarily focused on software, it can feel lonely. A software event is an opportunity to meet likeminded individuals — people who really care about the craft of software and improving themselves. Events are a good chance to make connections.

Keynotes

Many conferences, especially the larger ones, will include a keynote talk. Some include more than one. A keynote talk sets the tone for a conference and is usually related to the theme of a conference. For a multitrack conference, no other talks are usually scheduled during the keynote.

TIP

For a major conference held by a large software company such as Microsoft or Apple, the keynote is where they'll often make major announcements of new products and features.

Conference session tracks

The primary draw of conferences are the session tracks. A *track* is a set of talks, typically organized around a theme. Smaller conferences may have only a single track, while larger conferences may have a large number of tracks. A user group meeting or a meet-up may only have a single talk.

Depending on a conference, a session can range from 30 minutes to 75 minutes. Many of them end with some time for audience questions and answers (Q&A).

WARNING

If you participate in a Q&A, it's considered rude to simply use that time to make a statement. Make sure that your question actually ends with a question mark.

Sometimes presenters ask to follow up with you after the talk to have a more in-depth conversation. Don't assume that it's because they don't want to answer in a public forum. Typically, Q&A lasts only 5 to 15 minutes, and there isn't always enough time to fully understand a question and formulate an effective answer. You can meet up with presenters directly after the talk (though you should let them at least get their laptop and things off the stage) or ask when they may be able to grab a cup of coffee during the conference if it's a multiday conference. You can also ask whether there is an asynchronous way to follow up with them that they would prefer.

REMEMBER

Conference presenters are typically attending the conference to learn something as well; they aren't only there to present, so be respectful of their time as well.

Hallway tracks

At most conferences, the sessions are very valuable if it's a topic you want to learn about, but just as valuable is what's known as the *hallway track*. The hallway track isn't one you typically find on the conference schedule. It's a term that people coined to refer to the informal conversations you have in the hallway in between sessions, during lunch, or when you skip a session.

Sometimes you'll come across a world expert in a topic having a casual conversation with a few attendees, and you'll be welcomed to join in. These hallway tracks can often be even more educational than attending a talk because you can engage in a conversation instead of just listening. They also may lead to lifelong friendships and interesting collaborations.

TIP

Meeting new people at a conference can be intimidating, especially if you're an introvert. It may help to know that most people feel this way. If you draw up the courage to introduce yourself to someone you don't know who is alone, they may feel relieved. If you're an extrovert, look for opportunities to draw people who are alone into a conversation in a low-pressure manner.

After-hour conference events

Many conferences include after-hour events. Some of these events can be quite lavish depending on the size of the conference.

WARNING

Often, these events include loud music and alcohol, so be aware if that's not your thing.

Many conferences try to be more creative and inclusive with their attendee events. For example, one conference rented a table tennis place with a large number of tables.

A respectful professional environment

At most events, you can expect a respectful, inclusive, and professional environment. This environment is conducive to networking, socializing, and learning.

Occasionally, conferences fail to live up to that expectation. Because of that, many conferences adopt and enforce a code of conduct. A *code of conduct* outlines behavioral expectations of the participants in a conference. More importantly, it communicates to folks who are often the targets of harassment that the conference takes harassment seriously and is not welcome at the event.

Becoming Familiar with GitHub Events

Given that this book is about GitHub, I'd be remiss not to include some information about GitHub's own events. GitHub hosts and sponsors events throughout the year.

GitHub Universe

GitHub Universe (https://githubuniverse.com/) is the flagship conference for GitHub. It's held yearly in the city where GitHub's main headquarters resides: San Francisco, California. The conference is usually held in the fall around October or November.

As GitHub describes it,

GitHub Universe is a conference for the builders, planners, and leaders defining the future of software.

This conference is where GitHub typically makes its biggest announcements of the year during the keynotes. It attracts well-known speakers from prominent software companies.

In 2022, the conference was held hybrid, with sessions, booths, and demos in person, and the main session track also streamed live.

GitHub Satellite

The GitHub Satellite conferences (https://githubsatellite.com/) are an offshoot of GitHub Universe. They bring a GitHub universe-style conference to locations around the world.

Held once a year, past Satellites have been held in places such as Berlin, Tokyo, and London.

GitHub Constellation

GitHub Constellation (https://githubconstellation.com/) is a series of small community events held multiple times a year around the world. These events focus on the local community and often feature speakers local to the area. They are typically free and occur over one or two evenings. They're not all-day conferences like Satellite and Universe.

Git Merge

Git Merge (https://git-merge.com/) is a conference sponsored by GitHub but focused on the Git version control tool and the people who use it every day. As GitHub puts it,

Through technical sessions and hands-on workshops, developers and teams of all experience levels will find new ways to use, build on, and scale Git.

The conference features a preconference hands-on day of workshops focused on a range of Git topics. This conference is a great one to learn more about Git and to improve your Git skills.

Speaking at Events

A lot of developers have the misconception that speaking at conferences is only for experts who have been in the industry for years or that only big extroverted personalities speak at conferences. I'm here to tell you this is not true.

Everyone has a story to tell

Whether you realize it, somewhere inside of you, you have an interesting story to tell — even if you're relatively new as a software developer. One thing all these events have in common is they're better off when they have a diverse set of speakers with diverse viewpoints. For example, it's common for experienced developers to lose their "beginner's mind" when working with a technology for so long. A talk by a beginner about their struggle to learn a particular technology is often eye-opening and just the kick in the pants experienced developers need to make it better.

Benefits of being a speaker

Being a speaker offers a lot of benefits. The main one is that it's a forcing function to spend time deeply learning a topic. If you plan to give a talk on a subject, it's a good idea to research it beyond what you already know. And teaching a topic to others is a great way to solidify your own understanding.

Not only that, it's a great way to receive feedback on your ideas. Often, after giving a talk, someone will have a unique insight to share that improves upon your ideas. You wouldn't have received that feedback without putting your ideas out there.

Another benefit is the exposure and networking opportunities that being a speaker entails. Because your badge will have the word SPEAKER on it, people are more likely to want to meet you and talk to you. When you've spoken at a few conferences, it gets your name known. When you apply for a new job, people may give you more opportunities because they've heard you speak.

And a few perks come with being a speaker. Often, conferences have special speaker-only events where you can get to know other speakers. This event often leads to great networking opportunities and friendships because it's easier to remember people in a smaller setting. Also, as you speak at more and more conferences, you may see some speakers at multiple conferences and even become friends. It helps when you go to a conference to already know some of the other people there because you've spoken together before.

Finding Funding for Events

Some of these events and conferences can get expensive. Jamstsack can cost upwards of $1,000 — not to mention the transportation and parking to get to the conference and the hotel and flight costs if you don't live near the venue.

Many software companies pay to have employees attend a conference if the conference is a valuable learning opportunity. If you work for a software company, it doesn't hurt to ask. To make your case stronger, explain how the things you learn at the conference will improve your performance at work.

Larger software companies may also be sponsoring events and need volunteers to attend and help run the booth. Some companies even offer a stipend for attending conferences as a perk and part of your compensation package.

Another way to fund your trip is to apply for a scholarship. The Grace Hopper Celebration is a large conference with more than 18,000 attendees and celebrates women in computing. With technical tracks as well as tracks focused on diversity and inclusion, this conference is typically held over three days and moves around the United States for the venue. The Anita Borg institute that puts on the conference also offers scholarships that you can find at https://ghc.anitab.org/attend/scholarships/. These scholarships typically include airfare, hotel, transportation costs, meals, and a ticket to attend. They focus on students and faculty for this particular scholarship.

Most conferences waive the price of the conference ticket for speakers. Many conferences also cover hotel and travel to the conference. And some conferences even offer an honorarium on top of expenses. These benefits completely depend on the conference. As you improve your technical speaking skills and speak at more conferences, you can also turn this part of technical expertise into a career. Whether you decide to become a developer advocate (which often includes giving talks, making videos, live streaming, and educating developers on new tools or concepts) or you remain an engineer but commit to sharing your expertise like Kelsey Hightower (https://twitter.com/kelseyhightower), leveling up your speaking skills can bring a lot of perks and respect amongst the technical community.

Furthermore, some conferences ask for volunteers to help run the conference, which also tends to come with some perks such as a free ticket so that you can enjoy the rest of the conference when you're not working one of your shifts. The Grace Hopper Celebration calls it the Hopper Program, which you can find at https://ghc.anitab.org/get-involved/volunteer/hoppers/.

REMEMBER

If you've found an event you're particularly excited to attend, before you shell out the thousands of dollars from your own pocket, ask around! Ask your network whether anyone knows of any scholarships, apply to be a speaker or volunteer, and ask your company or school what resources are available to you!

7 The Part of Tens

Chapter **18**

Ten Ways to Level Up on GitHub

B ecoming an expert on GitHub is not a quick task. First, you need to master a lot of specific features of GitHub. Knowing how to create a pull request or link issues to a project board with automation for effective project management is one aspect of this expertise.

It's also important to begin to master specific areas in software engineering as a whole to be able to effectively contribute to projects in meaningful ways. Furthermore, becoming an effective community member is another way to become a GitHub expert.

This chapter briefly describes ten ways you can level up in each of these areas so that you can be successful on GitHub.

Trial and Error

One the best ways to learn anything is to just try. When you try, you always learn. If you try and fail, you learn what not to do and gain insight into how something works. When you succeed, you learn what to do next time! Learning how to use

GitHub and how to code is no different. In fact, learning by trial and error in the tech field is even easier.

TIP

GitHub provides you with unlimited public and private repositories, which means you can try all kinds of things without ever spending a dollar! I highly recommend getting onto GitHub.com and creating a new repository. From there, make a README.md file that you can easily modify and work with. This is the "Hello World" app of a GitHub repo. You can create issues and pull requests that modify the README.md file just to see how they work.

To learn more about collaboration on GitHub, invite your friends to be collaborators on your repo or make your repo public and send them the link to it. Ask them to comment on issues and create and review pull requests. Ask your friends to make their own public and private repo where you're a collaborator so that you can try all the GitHub features from the perspective of a contributor. There is no harm in trying things.

After you finish exploring, you can always delete the repositories, if you want. Keeping some around may be useful for the inevitable time when GitHub releases new features that you want to try on a dummy repo.

GitHub Help Docs

The GitHub help docs are extensive and detailed. You can find them by going to https://support.github.com. The help docs can be extremely useful if you know exactly what you want to get done. With 38 categories that each have anywhere from 1 to 70 docs, hundreds of pages describe every single feature of GitHub with images and cross-referenced links.

At the top of the GitHub Help page (where all the docs are located) is a search bar where you can type a query, and docs related to that query appear. In fact, just like with Google, results appear as soon as you start typing. If you press Enter or click the magnifying glass button, a page with search results appears. On this page is a notation next to the doc title with the category that it's a part of. I recommend paying attention to that notation because it will help you better navigate docs in the future. If you understand how GitHub categorizes the docs, you may be able to find what you're looking for faster next time.

If you click a specific doc, you see the documentation page. On the right side of this page is a list of Article versions. The documentation defaults to GitHub.com documentation, but GitHub also offers a slightly different version to enterprise users. Clicking the different article versions displays the docs specific to that version of GitHub, ensuring that you have the most accurate information.

TIP

Extremely security conscious companies are more likely to use GitHub Enterprise, as it allows them to have the interface and functionality of GitHub while keeping their code and data secure on their own servers. This ability can be important if you have proprietary code that you don't want anyone, not even GitHub, to have access to. If you're planning on open sourcing your code (which a lot of large companies, such as Microsoft, do a lot of) or you just want your code to be private, storing on GitHub.com is typically good enough.

At the bottom of each documentation page you typically find a section called Further Reading. This section lists documentation pages that may be relevant to what you're trying to do. Below this section is a Contact a Human button. If the documentation isn't helping you resolve your issue, click this button to go to a page where you can contact GitHub. The Contact page has an extensive list of all the ways you can try to help yourself first, but then provides a quick and easy message section for you to ask a question. On the left side of the contact page you can also find other reasons for contacting GitHub, such as reporting abuse, reporting content that isn't appropriate, reporting copyright infringement, privacy concerns, and portals for both premium and enterprise support, if you've purchased those plans. You can also directly access this page at `https://support.github.com/request`.

WARNING

This Request page is not where you should submit a question or concern about a specific repository unless you're reporting abuse, content, or copyright infringement. If you have a problem getting code to run or have a question about a specific repository, you should ask in an issue directly on that repository home page on GitHub.

REMEMBER

Docs are your friend. There is a lot to learn and software changes quickly; expert programmers are experts not just because they know a lot, but mostly because they know how to figure something out — and that often means knowing how to find information.

GitHub Skills

GitHub has invested in a team of folks dedicated to helping novices learn how to use GitHub. And though documentation can help when you know exactly what you're looking for, sometimes it can be hard to follow still images. That's why GitHub brings you Skills, which are absolutely incredible. Skills are step-by-step, guided tutorials where you actually create a real repository on GitHub and perform real actions that you would normally. The friendly GitHub Skills Bot guides you through the tutorial, creating pull requests for you to review or issues for you to close out.

To get started with Skills, go to `https://skills.github.com` and browse a number of courses that may be useful for you to get an in-depth introduction into everything GitHub.

On the Courses section, on `https://skills.github.com`, you can find 14 (maybe more by the time you read this book) courses categorized into topics such as First day on GitHub, Automate Workflows with GitHub Actions, and Code Security and Analysis. With some of the most commonly used features on GitHub being highlighted, you can learn to review pull requests, manage merge conflicts, and make your open source project stand out among the millions of projects on GitHub.com.

Clicking a course takes you to a detailed page where you can learn more about that course and join it. Often times, Skills wants to create a public repository on your account. Don't worry; you can always delete it after you've finished the course, but there is also nothing wrong with showing the world that you take learning seriously. When you start a Skills course, you're taken to a Template repository. Clicking Use This Template, as shown in Figure 18-1, to clone the repository in your own account and the GitHub Actions automates the course experience for you.

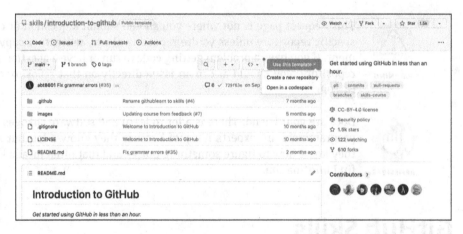

FIGURE 18-1: Using the GitHub Skills template repository.

A typical Skills course has the bot open an issue in the repo that it created with a description of what you should do. All you have to do is read through the issue and follow the instructions!

Open the repo in a new tab so that you can refer to the instructions without losing your place in what you're trying to accomplish.

TIP

When you're just starting to get introduced to GitHub, Skills is a great way to experiment with features and learn specific workflows. It's essentially trial and error, but with a helpful guide.

GitHub In-Person Training

GitHub is tool for more than 30 million developers and over 2 million organizations. From novice coders just learning to create their first GitHub Pages website in Markdown to Microsoft bringing you VS Code, GitHub's goal is to support more people building more software. As such, GitHub offers trainings where a GitHub expert helps better equip your team to use GitHub.

With eight different focus areas, plus a customized training in case you need something different for your team, GitHub not only guides you through the details on how to use GitHub, but engages everyone in the fundamental workflows and techniques to using it effectively. The focus areas are

>> Workflow consultation

>> Implementation

>> Admin mentoring

>> Train-the-Trainer

>> Migration

>> API consultation

>> InnerSource

>> Services account engineering

For a full description of each of these areas, as well as what you can get with a customized course, visit `https://services.github.com`. This service is not free. You can submit an inquiry and a GitHub representative will reach out with more information about costs.

If you're a novice, you may not need the full-force of a GitHub trainer. However, it may be something for you to suggest to your company if you need to start using GitHub as part of your job. In-person training can also be useful if you start a company and want everyone on your team to not only use the tool, but to use it effectively.

Project-Specific Documentation

One of the best ways to learn about a specific open source community and a specific technology is to reach the project-specific documentation. When you are both new to GitHub and new to a technology, it can be effective to read through the documentation most relevant to your interest and to explore the behaviors of other members of the community by reading through issues and pull requests.

The VS Code project is a good example of a project with good documentation and enough community engagement to understand behaviors. If you go to https://github.com/microsoft/vscode, you can find a well-written README with links to where you can submit bugs and feature requests, how to contribute to the project, and where to engage with the core developers.

When you're first starting to learn, following the How to Contribute documentation (for example, VS Code has a wiki page for it at https://github.com/Microsoft/vscode/wiki/How-to-Contribute) can help you get your code up and running on your local machine. Even if you don't end up contributing directly to the project, following this documentation can be a good learning exercise.

REMEMBER

Some projects are strict on their style and contributing guidelines. For example, the VS Code project has guidelines on style found at https://github.com/Microsoft/vscode/wiki/Coding-Guidelines. Having code that is consistent in style means that errors are less likely to occur. For example, if you always use PascalCase for type names and camelCase for function names, you can quickly identify when someone accidentally referred to a type instead of a function in their code.

It is also important to know the requirements for contributing code to the open source project. For example, on VS Code, you must first sign a Contributor License Agreement before you open a pull request on the repository.

Looking through open and closed pull requests can also be good practice when you're trying to learn about the specific community. Understanding what kinds of errors are common in the particular code base or what kinds of changes the core team typically requests can help you learn about the technology as well as the specific project. Successful open source projects have pull request reviews with substantial information. For example, on one of the pull requests for the VS Code project, you can see user ramya-rao-a not only lets user grunxen know that alerts shouldn't happen, but also provides a suggestion on how to fix this bit of code, as shown in Figure 18-2.

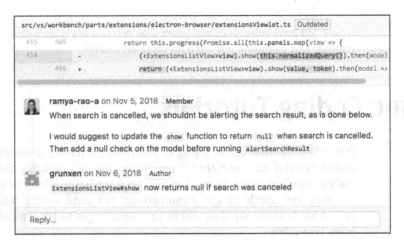

FIGURE 18-2:
Effective pull
request review
feedback on the
VS Code open
source project.

External Community Places

Some open source projects have places other than GitHub where more general conversations can happen. These places can be a really great way to learn without having to explicitly open an issue or pull request. You can find these extended communities many places online, some with a more interactive community while others are more feedback-driven.

The VS Code project, for example, calls out the StackOverflow questions that are tagged with visual-studio-code. If you have questions about VS Code, it's best that you first search the questions already asked. If yours is unique, make sure you use the appropriate tag when writing it.

This project also suggests giving feedback on Twitter, asking questions, or directly giving feedback to the @code alias. Giving feedback is also a good way to get updates on the project.

The Sentry (application monitoring) open source project, on the other hand, uses the GitHub Discussion Forums, found at https://github.com/getsentry/sentry-javascript/discussions. This forum is a great place to search for answers to your question or ask a unique one, if it hasn't been asked. As you become more expert in using Sentry, you can start contributing answers.

Sentry also has a public Discord that you can join. Just visit https://discord.gg/sentry and you will be invited to join the public Discord server where you can ask questions and generally follow the discussion around Sentry, any of its specific integrations or SDKs, or just general chat "around the water cooler." This resource is great if you're learning on your own. A Discord server can be a virtual office where you have colleagues and peers that you can ask questions to in a lightweight manner (for example, not open an issue on GitHub).

Online Coding Tutorials

Depending on what you're trying to learn and contribute to on GitHub, an online coding tutorial may be right for you. Now, you can probably find thousands of coding tutorials online from a random person posting a blog to a YouTube video showing you exactly how to do something, but some coding tutorials also offer you an in-browser sandbox where you actually get to try the code it's trying to teach you about.

If you ever want to create a GitHub app or integration, it may behoove you to know a bit about Ruby because GitHub is built using Ruby. The online tutorial https://try.ruby-lang.org/ gives you a step-by-step guide with an editor and an output window right in your browser.

TIP

While the Ruby tutorial gives you a handy Copy button so that you don't have to type all the example code every time, it can be useful to go through the motions of typing to slow down and better learn what you're reading. I recommend typing the code, even if it seems simple enough, so that you get more practice.

If you're making a web application, you may want to get some practice with JavaScript, HTML, and CSS. https://jsfiddle.net is a great place to start, giving you an editor for each language along with an output window. It has also been enhanced to have boilerplate starter code for commonly used languages, such as React, TypeScript, and CoffeeScript.

Another interactive sandbox tutorial website I particularly like is the Try .NET site at https://dotnet.microsoft.com/platform/try-dotnet. Before diving in to the intricacies of the .NET Framework, you can get your hands dirty in a safe environment right in your web browser. With snippets that you can try, tutorials you can follow, or simply an open playground to experiment, this site can offer you support at any stage in your career.

Online Courses and Tutorials

Online courses are becoming more popular with the development of better tools for engaging with learners. Two of the largest platforms for online courses are Coursera and EdX. These platforms partner with universities and large companies to provide in-depth education and sometimes certification or even degrees.

Both Coursera and EdX offer specializations with multiple courses. For example, there is a Ruby on Rails Web Development Specialization on Coursera that was created by Johns Hopkins University. Meanwhile, Microsoft offers a professional certificate in Introduction to Computer Science, which has three courses.

Furthermore, some universities offer degrees, such as the University of London and its bachelor of science in computer science degree on Coursera or University of California, San Diego's MicroMasters in data science on EdX. These programs often require an application and admission into the specific program — not just anyone can join.

For more lightweight options, sites like Udemy and Pluralsight offer more than 100,000 online courses at low prices or sometimes even free. These courses tend to be more focused on video lectures and tutorials rather than extensive coursework, but they're really great options for when you're first getting started and just want to get an introduction into something.

Khan Academy is another great place to find always free courses. Khan Academy supports learners as young as 2 on its Khan Academy Kids app (`https://www.khanacademy.org/kids?from=lohp`) to adults looking to learn something new from computing to studying for the LSAT to entrepreneurship and growth mindset! This great resource offers curated courses that are more lightweight than Coursera or EdX.

It is also important to not forget the power of YouTube. Though videos on YouTube aren't curated and aren't held to any standards, you can find a lot of really amazing content. For example, The LearnCode.academy channel has a GitHub Tutorial for Beginners video with almost two million views. Siraj Raval is also a great channel to follow as he gives tutorials, but also explains the history of certain technologies and how they actually affect our world.

Blogs, YouTube, Twitter, TikTok, and Other Social Media

If you're looking for updates on certain technologies or products or a quick answer to something, it can often be fruitful to follow the right people on Twitter, YouTube, Mastodon, Twitch, TikTok, or whatever social media platform you're into and subscribe to the right blogs. I can't tell you who the right people are for you to follow, because each person reading this book may have different interests when it comes to learning and contributing on GitHub, but I can give you some tips!

First, recognize that people who post on social media and who blog are usually only representing their own ideas and are not representing a company or being held to any company standards. This means that you should take what they say with a grain of salt and recognize that they are just people, saying what is on their mind, and probably also still learning and developing (as we all are always learning and developing).

When searching for Twitter accounts to follow, you can start with accounts representing the technology you're interested in learning about. For example, following @GitHub may be a good idea because this account not only posts updates to GitHub, but also often highlights other tech, people, or events related that may be of interest to you. This start to give you ideas on other people to follow. For example, GitHub recently tagged Finn Pauls (@finnpauls), who is an engineer and product designer who shared some open source projects, so he might be a good resource of information.

Advice on which blogs to follow is similar to the advice on Twitter. I recommend starting with some of the blogs that may have the most relevant information — for example, the GitHub blog at https://github.blog gives you up-to-date information about GitHub features. Another great resource is Dev.to, where you can search any keyword and get a number of different perspectives and pieces on it. For example, https://dev.to/search?q=github results in a lot of lists of things you can do on GitHub that you may not have known about. It can be a great start to then look into the GitHub docs for more in-depth information. Finally, you can also just enter a keyword in your favorite search engine, and various blogs, tutorials, and videos are likely to pop up.

Community Forum

I also recommend the GitHub Community Discussions. especially for getting help about GitHub. Found at https://github.community, this forum tends to have hundreds of people actively online at any given time. The GitHub community is worldwide, and people from all different technological backgrounds are typically eager and willing to help.

Many threads contain topics specific to GitHub, such as how to use Git and GitHub, GitHub Pages, and the GitHub API. But this community forum offers even more than GitHub-specific information. As the number one place for developers, GitHub also provides a space for the community to learn from each other in general programming concepts and projects overall.

You might also be interested in the GitHub Education community forum found at https://education.github.community focused on how to use GitHub in the classroom. This resource is great if you're a student and you'd like to convince your teacher to use GitHub or if you're a teacher looking to teach your students.

The GitHub Community Discussions is a great place to ask your generic questions that wouldn't be appropriate to ask on a specific repository. The GitHub community is vast as it is a large portion of the developer community. Almost anywhere you go in the developer community, you are sure to find someone who can help you or at least point you in the right direction. The most important thing for you to remember is to ask questions, be respectful, and remember to give back with your knowledge as you begin to learn.

Chapter 19

Ten Ways to Improve Your Development Workflow

Working on software is tedious at times. Writing code is laborious and requires a ton of steps and intense concentration. On top of that, a lot of tasks have to occur during the coding process, such as running tests, creating mock-ups, and tracking progress.

Any tools and techniques you can use to help streamline and improve your development workflow not only saves you time, but can improve the quality of your work. It's always worth spending time periodically looking at ways to improve your development workflow. In this chapter, I cover ten ways you can improve your development workflow.

Drafting Pull Requests

Chapter 8 discusses creating pull requests when you're ready to have code reviewed and merged into the main branch of a repository. But that's not the only way to use pull requests. In fact, GitHub employees have long stated that creating a pull request is the beginning of a collaborative conversation. Sometimes it may be appropriate to create a pull request even before you've pushed code. It's possible to do by creating a branch on GitHub and then creating a pull request from that empty branch.

Or maybe you do have some code to push, but you know it's incomplete. You just want to gather some feedback on your progress without alerting code reviewers that they should give your pull request their full attention with a detailed code review. This is where draft pull requests come in handy. To draft a pull request, click the Create Draft PR button, as shown in Figure 19-1.

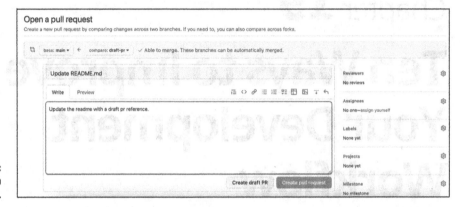

FIGURE 19-1:
Drafting a pull request.

This creates a pull request in draft mode. If you have a CODEOWNERS file (which I cover in Chapter 11) in your repository, a draft pull request doesn't notify those reviewers until the pull request is marked as ready for review by clicking on the Ready for Review button, as shown in Figure 19-2.

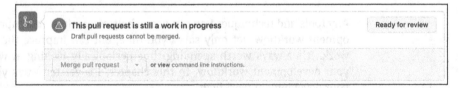

FIGURE 19-2:
The Ready for Review button.

Git Aliases

Chapter 6 introduces the concept of Git aliases. Git aliases are short shell scripts that extend Git and automate common tasks. Chapter 6 includes a Git alias for moving commits from one branch to another.

Git aliases can automate tedious clean-up tasks. For example, the following alias deletes all branches that have already been merged into the target branch. If no target branch is specified, then it assumes the main branch.

```
[alias]
bclean = "!f() { git branch --merged ${1-main} | grep -v "
            ${1-main}$" | xargs git branch -d; }; f"
```

Aliases can be combined together. For example, before you clean up branches, you may want to switch to the target branch and update that branch from the remote server first.

```
[alias]
bdone = "!f() { git checkout ${1-main} && git pull --rebase &&
            git bclean ${1-main}; }; f"
```

Note how the bdone alias makes use of the bclean alias. You might use the bdone alias right after you push a branch that then gets merged by someone on GitHub. com. After the branch is merged, you can run git bdone, and the alias switches you back to main, runs a git pull, and then runs the bclean alias.

TIP

Adding aliases you find from elsewhere can be a pain, though. This post describes a quick way to include a bunch of aliases into your Git config file: https:// haacked.com/archive/2019/02/14/including-git-aliases/.

Run Tests Automatically

Writing automated tests for code, such as unit tests, is an essential skill for professional software developers. It helps improve the design of code and provides a safety net when making changes to code.

When working with a code base with a lot of tests, it's not uncommon to forget to run tests often. And even if you do run tests often, it's a distinct step: Make some changes and then run a command to run your tests.

A great way to improve your development workflow is to automate the test runs. Many tools automatically run tests when your code changes. Here's a short list of tools for various platforms:

>> NCrunch for .NET

>> Guard-test for Ruby

>> Tsc-watch for TypeScript

>> Cargo-testify for Rust

>> Wallaby.js for JavaScript

Many of these tools are smart about only running the tests affected by the changed code. That way, every change doesn't end up running the entire test suite.

Take Breaks

One simple effective method for improving your workflow, albeit one that is neglected by many developers, is to take regular breaks. Writing code may feel like a sedentary task, but any activity done over an extended period of time can put a lot of stress on the body. The human head is pretty heavy (some heavier than others). Holding it upright can put a lot of stress on the back and neck. According to DataHand, a maker of ergonomic keyboards:

At the end of an average eight-hour workday, the fingers have walked 16 miles over the keys and have expended energy equal to the lifting of 1 1/4 tons.

Taking regular breaks to stretch your hands, arms, and back can help you remain healthy and thus more productive.

However, that's not the only benefit of taking breaks. Many people are fans of the Pomodoro Technique developed by Francesco Cirillo in the late 1980s. This technique breaks work down into intervals, traditionally 25 minutes, with a short break of around 3 to 5 minutes in between. Each interval is known as a *Pomodoro*, Italian for tomato. Why a tomato? Apparently, Cirillo used a tomato-shape timer in college.

After four pomodoros, you take a longer break (15 to 30 minutes). The benefit of this technique is not only to enforce that you rest your body, but it has the added benefit of helping you maintain focus. The general idea is that during a pomodoro, you are intently focused on work. You generally close all other distractions. You can use some of the longer breaks to do things like check emails and browse the web, if you need to. But during the 25-minute pomodoro, you should be intently focused on work. Practitioners swear by the increased focus and flow the technique encourages. You can find many examples of pomodoro timer applications on the web.

Prototype User Interfaces

Whether it's a web or desktop application, building a user interface can be very time-consuming. And it's difficult to know how usable an interface will be until you put a human in front of it to try it out.

One tool that saves a lot of time when building an interface is a rapid prototyping tool, such as Balsamiq or Invision. This is by no means an exhaustive list of such tools.

The benefit of these tools is they make it possible to build mock-ups of a graphical user interface (GUI) in a short amount of time. Some tools even make it possible to add a bit of interactivity so that you can put the interface in front of a person and run some informal usability tests. You can get a lot of valuable feedback by simply asking people questions like "How would you accomplish a task on this screen? What would you click next?"

TIP

Making changes to the interface to respond to such feedback is much faster than if you had written a bunch of code. Once you've run through a few iterations with the mock-ups, you can build the actual GUI with more confidence that you're on the right track.

Scaffold Apps with Yeoman

Starting a new application from scratch can be time-consuming. A lot of boilerplate code goes into setting up a real production application, and the boilerplate is different depending on the type of app and what the app does.

Yeoman is a tool for scaffolding modern web apps. To install it, run the following command:

```
$ npm install -g yo
```

This command adds the yo command to your machine. Yeoman works with generators, which are essentially plugins to the yo command, that add support for a given project type. A huge ecosystem of generators is out there.

For example, suppose that you want to build an extension for Visual Studio Code (VS Code). You would start by installing the generator for VS Code:

```
$ npm install -g yo generator-code
```

To run the generator, you run the yo command with the generator name:

```
$ yo code
```

The generator prompts you to answer some questions about the project to generate, such as specifying a project name. At the end, the generator takes your answers and scaffolds a project folder with a working VS Code extension.

Chrome Web Developer Tools

If you develop browser-based applications for the web or for the desktop via Electron, no tool is probably more useful than the Chrome Web Developer Tools. To launch these tools, press ⌘-Option-I on the Mac or Ctrl+Shift+I on Windows.

You can also launch the developer tools by right-clicking an element of any web page and choosing the Inspect Menu option. When the developer tools open, select that element in the Elements tab.

The Elements tab of the developer tools allows you to explore and manipulate the DOM. You can also manipulate the CSS. This provides a nice way to debug CSS problems because it gives you instant feedback on your CSS changes.

The Console tab lets you run JavaScript commands in the context of the current web page.

The Sources tab lists all the scripts that are loaded in the context of the page. This list can be eye-opening when you go to a website you use often and look at this tab. A given web page can have a large number of scripts running.

The Network tab shows all the network requests that were made to render the page, the size of the requests, how long they took, and when they happened in relation to each other. This information is useful for debugging issues where a large request is causing delays in rendering a page.

The Performance and Memory tabs are useful for profiling execution time and memory usage of a page.

StackOverflow

StackOverflow.com is a question and answer website for developers. Since its creation in 2008, it's had a huge impact on the developer community. A big part of its popularity is due to the gamification techniques it employs to maintain high-quality questions and answers.

For example, questions and answers can be up voted and down voted. Answers that receive the most up votes are displayed directly under the question so that people who find the question later don't have to wade through a ton of answers to find the best answer. If the poster of a question accepts an answer, that answer floats to the top (unless the poster also answered the question), regardless of the number of up votes. This setup sometimes causes a situation where a better answer with more up votes is displayed before the accepted answer.

Asking questions, answering them, and having questions or answers voted up all contribute to your reputation points. As your reputation points increase, you gain more privileges on StackOverflow, such as being able to edit questions and answers to collectively improve them like a wiki.

If you're stuck on a programming task, StackOverflow is often a good place to start searching for an answer to your question.

Code Analysis Tools

A wide range of tools can analyze code for potential problems and potential improvements. When used properly, these tools can save you a lot of time and headache.

Linters are a class of tools named after a Unix utility named Lint that analyzes C code to flag bugs, style errors, and potential problems without having to run the code. While Lint is the original tool, many linting tools exist for different programming languages, such as JSLint for JavaScript and ruby-lint for Ruby.

Static analysis tools are similar to linters but work against statically typed languages. These tools take advantage of type information in the source code to find issues in the code that aren't syntax errors (which the compiler would already catch) but may cause problems down the road. For example, static analysis tools can flag code that may exhibit poor performance in certain situations. Examples of static analysis tools include FxCop for .NET and Coverity for Java.

Some tools surface metrics about your code that may be correlated to quality, such as measuring cyclometric complexity. Cyclometric complexity refers to the number of different execution paths through a piece of code. A method with a very large cyclometric complexity can be hard to understand and prone to bugs.

Some tools, such as Code Climate Quality, are available in the GitHub Marketplace and can perform automated code reviews looking for common problems in code. This tool can identify files and sections of code that are changed frequently. Frequent changes often indicates that the code may have quality issues. Understanding where your code churns often helps you focus your attention on changes to that code.

Project Boards

Project boards are a useful way to visualize the progress and tasks for a project. Chapters 3 and 4 walk through setting up a project board along with project automation for a repository. For a given repository, a project board serves as a common source of truth. Everyone can look at a project board and have a good idea of the overall progress at a glance.

However, you may consider using individual project boards that are not associated with a single repository as a means of managing your overall set of tasks on GitHub. You can go to https://github.com/dra-sarah?tab=projects to create a new personal project board.

Project boards come with a limited set of automations. To really customize your workflow, you may want to create a GitHub Action (see Chapter 14). With GitHub Actions and the GitHub API, it's possible to create project board automations for nearly any workflow you can think of.

GitHub project boards are not the only option for a Kanban-style board that works with GitHub. Trello (see Chapter 14) is another option. A few other options with deeper integration with GitHub include ZenHub, waffle.io, and HuBoard.

Regardless of the one you pick, a good project board is a helpful addition to any software developer's workflow.

Chapter **20**

Ten Tips for Being an Effective Community Member

As of June 2022, more than 83 million people are on GitHub. That's a lot of people. It can be difficult to stand out among such a large sea of people. But the truth is, most people starting out on GitHub don't really understand how to be effective. They don't understand the rules of the road to being a great member of the community.

In this chapter, I compile ten tips that will help you be a wonderful community member — the type of person that every maintainer is excited to have involved with their repositories.

Be Respectful and Kind

Maintaining a repository can be a frustrating affair — especially when it's a popular repository and you're volunteering your free time to an open source project and you're about to answer the same question for the hundredth time.

It's understandable that you may be very curt to the next person who asks the same question. It's a waste of your time, and they didn't do their due diligence to search to see whether the question was already answered.

WARNING

Resist the temptation to lash out. For that person, it may be the first time they've ever created an issue. They may not have read this chapter yet and learned how to be effective on GitHub. Your response sets the tone for their experience of your project and, perhaps, of GitHub as a whole.

Be respectful and kind, and you may win over someone who will become a lifelong contributor. And even in the face of rudeness, remember that we all have bad days. You've probably done the same in the past yourself. The Internet can be very impersonal when communicating in writing. Sometimes a little kindness at the right moment reminds people that a human is on the other side and there's no need to be rude.

REMEMBER

But to be clear, killing them with kindness is fine for the occasional rude behavior, but you should not have to tolerate abusive behavior.

Report Bad Behavior

Whether bad behavior is directed at you or others, it's important to the health of a community that you remain vigilant and report it. If you encounter or witness abuse on GitHub, report it at `https://github.com/contact/report-abuse`.

TIP

It can be difficult to know the difference between someone just being slightly rude and someone being abusive. GitHub's Terms of Service help spell it out. If you're unsure, know that GitHub's support people are well trained to draw this distinction and will not react in a knee-jerk fashion to your report.

Abuse isn't the only type of issue you may want to report to GitHub. You can report harmful content, privacy concerns, and copyright claims from GitHub's contact page at `https://support.github.com/tickets`.

Write Good Bug Reports

For a repository maintainer, few things are more frustrating than an issue that looks something like this:

The thagomizer isn't working, fix it! I tried it out, and it doesn't work.

This example may seem extreme, but vague bug reports that are not actionable are unfortunately quite common. It doesn't take too much effort to write a bug report that is helpful.

TIP

When you come across a bug report where you have some expertise, don't respond with "Works on my machine" with no other follow-up. Confirming that it works on your machine is actually beneficial, but it is more helpful to offer suggestions for what may be different between your machine and the person's who filed the bug report. For example, you may want to let the person know which version of the software you have installed or if you have any other setup that may affect the situation.

Sometimes, the best bug report is the one that isn't written. Before you create an issue, search the issue tracker to see whether someone else already reported the bug. If they did, you may want to add further details to the issue if you see anything missing.

Assuming you didn't find the issue, it's time to create a new issue. If the repository has an issue template, you should follow the template as closely as possible.

If is the repository has no template, the following is a good format to follow:

1. **Describe the observed behavior.**

   Try to be objective and communicate facts, not opinions.

2. **Describe what you expected to happen.**

   How did the observed behavior differ from your expectations?

3. **Describe detailed repro steps.**

   This step is the most important part of the bug report. Describe step by step how someone else can reproduce the problem. Try to make the repro steps as minimal as possible. When you encountered the bug, you may have taken ten steps to get there. But it may be possible to reproduce the bug in only five steps. Spending a little time to make sure that you remove any extraneous steps goes a long way into making a good bug report.

4. **Describe the repro environment.**

   This step is where you describe the environment where you reproduced the bug, such as the operating system and browser. While it's not necessary to try to reproduce the bug in other environments, repository owners are very appreciative if you do and report on the results.

Be Responsive

No matter how good your issue write-up is or your pull request code is, chances are the repository maintainers will have some follow-up questions. It's particularly frustrating for a maintainer to ask for more information only to be greeted by crickets. If you submit an issue, don't ghost on it. Make sure you make time to follow up and respond to questions from the maintainers.

REMEMBER

And the shoe fits both ways. If you are a maintainer, try and be responsive to people who submit issues and pull requests. In some cases, using an automated response to an new issue or pull-request is appropriate if your repository is particularly busy. For example, you could use a Probot app to automatically respond to new issues and pull requests with a note letting them know you plan to look at it but that it may take some time.

Being responsive doesn't necessarily mean you take care of everything right away. It means that you set expectations right away. On either side of the comment, whether you're a contributor or a maintainer, the only way for the other person to know when to expect changes is if you tell them. Working on remote, asynchronous projects depends heavily on communication. When you're in an office you can see whether someone doesn't come in to work for a while and know they are probably on vacation; on GitHub, you don't have that contextual awareness.

TIP

As an added form of communication, keep your profile status up to date if you do plan on going on vacation. For more detail, see Chapter 16.

Submit Pull Requests to Correct Documentation

The campsite rule states that one should leave a campsite in better condition than they found it. It's a good rule to follow not only with campsites, but also with documentation.

Good documentation is often the weak point with open source projects. Many OSS projects have very few volunteers, and much of their time is taken working on the actual code. This lack of support makes repository maintainers especially appreciative when someone comes along and contributes to the documentation.

Getting involved with improving a repository's documentation is also a great way to dip your toe into OSS. If you happen to find an error or something missing in a project's documentation, consider submitting a pull request to the project fixing the error. This contribution leaves the overall OSS ecosystem better off than it was before.

REMEMBER

One of the most challenging pieces of documentation to keep up to date is the Getting Started docs that contributors are meant to follow to first get the project set up on their machine. It's challenging because the maintainers rarely set up the project brand new, unless they get a new machine. As you're following the steps, don't be afraid to open a pull request on this documentation if you had to do something different. It can be especially helpful if instructions are different on Mac versus Windows versus Linux and you're on a machine that the project doesn't have documentation for yet.

Document Your Own Code

Documenting your own code goes a long way toward making it more accessible to others. Trying to use unfamiliar code can be challenging and time consuming. Good documentation can save people a lot of time and get them up and running with your code quickly. If only 12 people use your code and you save each one two hours of hassle, that's a full day saved!

Depending on the platform, many tools generate documentation from comments in code. Javadoc is a famous example for Java code. It requires that you comment public methods and classes with a standard format. By following the format, you can use the Javadoc tool to generate HTML files. JSDoc is another one for JavaScript.

In addition to code documentation, consider other documentation such as the ones I cover in Chapter 9. For example, every repository should have a README.md file that describes what the repository does. It should also have a CONTRIBUTING.md file that describes how to contribute.

Give Credit Where It's Due

WARNING

Most open source licenses require proper attribution if you make use of the code in your own project. If you use some source code from an open source project, giving credit where it's due is a legal matter and required by the license.

But to be an effective community member, credit doesn't end with attribution to comply with a license. In many situations, giving credit demonstrates that you are a classy person.

For example, if someone contributes a feature or bug fix, mentioning the person who fixed it in your release notes is a good idea. For example, GitHub Desktop makes a point to thank people using its GitHub handles and links to the pull request that contributed to a fix in its release notes at https://desktop.github.com/release-notes/.

Another great way to give credit is to mention the person who opened an issue that you may have fixed with a pull request. The pull request description will most likely link to the issue, but calling out the person who found the issue in the description is a great way to encourage others to continue to help find bugs.

Help Get the Word Out

Many open source projects are small and relatively unknown. If you find a project useful, help get the word out. It not only benefits the project that may get an influx of new users and contributors, but it benefits the people you tell who may need that very tool.

REMEMBER

You don't need a huge platform to help people get the word out. Maybe your Twitter follower count is relatively small. Don't let that stop you. The power of network effects can sometimes help a message really take off.

If you write blog posts or publish YouTube videos, you can also mention different projects, your impressions, and maybe even a tutorial on them on these mediums. The goal here is to help people and projects meet.

Be Proactive and Mentor Others

GitHub's community is growing rapidly. New developers are being minted every single day. What this means is a lot of beginners will be on the site. Over time, you will start to accumulate experience that would be very valuable to one of these beginners. Be proactive and offer to mentor others.

For example, if you maintain an open source repository and see that someone is struggling to make a contribution, offer to walk them through the process. Point them to resources and help them along. If you help two people become proficient contributors and community members and they each help two people down the road and so on, you could end up having a huge impact on the community.

Sometimes it can even be effective to jump on a video chat and help someone debug their code in real time! If you're comfortable with this approach, it can be a great way to meet new people and see for yourself how someone new approaches your project. It may provide insight into how to improve your documentation.

Contribute Outside of GitHub

Many open source projects have a lot of activity outside of GitHub. Contributing to the community can go beyond creating issues and opening pull requests. For example, as you gain expertise in a topic, consider heading over to StackOverflow. com and answering questions on that topic. Many open source projects have chat rooms associated with the project in Slack or Gitter. It's a benefit to others if you head over there and offer your ideas.

Talk to maintainers about other ways you can support their work. As you grow in your career as a software developer, you will pick up skills that are valuable to an OSS project. For example, you may help them figure out how to sign a package using Let's Encrypt. You may help them register a domain name and pay for it. You may help them navigate setting up a Docker container so that others can try out their project with less setup fuss.

Whatever it is, don't be shy in offering your skills in support of open source projects, especially those that you benefit from.

Index

Unspecified

. (period), 12
.git, 242
.gitconfig file, 107
.github folder, 159
.github/ISSUE_TEMPLATE folder, 178
.gitignore file, 39
.NET Framework, 286, 293, 297
.NET São Paulo, 268
/github subscribe command, 210
/github subscribe owner/repository, 208
/github unsubscribe owner/repo [feature]
    command, 211
:, 124
:art:, 117
:q! 115
:sparkle:, 142
:wq, 115
_config.yml file, 87
_layout/post.html file, 88
−A flag, 115
−m flag, 114
−−oneline flag, 105
−u flag, 131

Symbols

#greetings, 125–126
#ISSUEID format, 134
$, 10
$ git browse, 132, 134
$ github ., 119
$ open index.html, 112
* (asterisk), 83
@ (at symbol), 126
<commit-range> parameter, 107–108
<details> tag, 138

<new-branch-name> parameter, 107–108
<target-branch> parameter, 107–108

Numbers

404 error, 62

A

About page, 137
abuse, reporting, 172–173, 281
access, repository, 190, 192
accessibility, 19
accounts
    VS Code, 34
    menu, 27
    personalizing, 18–19
action buttons, Code tab, 43
Actions tab, repository, 42
actively reviewing pull requests, 77
admin mentoring, 283
after-hour conference events, 272
AI (artificial intelligence), 22, 239
Alerts insight, 199
aliases, 107–108, 131, 134, 292–293
All Things Open, 155, 270
animated gif, 139, 235
Anita Borg institute, 275
Apache License 2.0, 167
API consultation, 283
APP_ID, 239
appearance, 19
Apple
    WWDC, 270
    XCode, 226–228
apps, 236–243, 254–257
    code quality apps, 255–256

testing
  running automatically, 293–294
  software apps, 257
TestQuality app, 257
TextEdit, 69
text-message confirmation, 186
themes, 59
third-party access, 192
TikTok, 287–288
timeline, contribution activity, 265
titles
  creating issues for, 65
  modifying, 87
tokens, 23, 226
topics
  exploring, 152–154
  location of, 43
Torvalds, Linus, 266
tracking
  issues, 158
  repositories, 118–119
tracks
  hallway tracks, 272
  session tracks, 271
Traffic insight, 198
training, in-person, 283
Train-the-Trainer, 283
transferring ownership, 182–183
Trello, 211–216
  installing GitHub power-up, 211–213
  using GitHub power-up, 213–216
trending repositories, 154–155
triaging
  overview, 78
  own projects, 176–177
  reading through issues, 77
trial and error learning, 279–280
Try .NET site, 286
Tsc-watch, 294
tutorials, 286–287
Twitter, 287–288

two-factor authentication, 20, 223, 226
  of members on organization, 190
  on tiers, 186
TypeScript, 294

U

Udemy, 287
unblocking users, 173
University of California, 287
University of London, 287
unlabeled filter, 176
Up For Grabs website, 157
updates, VS Code, 34
up–for–grabs label, 157
upstream
  contributing changes to, 99–102
  fetching changes from, 98–99
user experience, 251
user groups, 268
username
  adding sections with, 87
  changing, 18–19
  forks having, 98
  mentioning in pull requests, 134
  setting up personal website with, 59

V

version control system
  Visual Studio, 223
  XCode, 228
version control systems
  Git, 8–16
    branching by collaborator, 14–15
    branching by feature, 15–16
    branching for experimentation, 16
    trying on terminal, 9–14
  GitHub, 8
VI editor, 115
VIM editor, 115
Visual Studio, 223–226

About the Author

Sarah Guthals, PhD, is a former engineering manager at GitHub and director-level of developer advocate at Microsoft and Sentry. She is coauthor of *Helping Kids with Coding For Dummies.*

Dedication

This book is dedicated to my daughter Ayla. Through one of the hardest years of my life, she has been a treasure and constant source of inspiration and resilience. I'd also like to dedicate this book to my close friends, family, and colleagues who have supported me in becoming who I am today, which has led to the opportunity to bring this updated book to you all today.

Author's Acknowledgements

I would like to acknowledge all of the hard work that went into making GitHub, all of the open source developers who share their passions on GitHub, all of the folks who build apps that integrate with GitHub, and every programming language, coding application, and effort to improve the collaborative nature of programming — without these, coding would not be as inspiring and fun. I would also like to acknowledge Phil Haack for originally co-authoring this book with me (his experience and wisdom is definitely a key feature) and Cecil for stepping in last minute to edit this book for technical accuracy. Finally, I want to give a huge thanks to Steve, Rebecca, and all the folks at Wiley for being understanding through intense challenges in my personal life and still making writing this book a great experience.

Publisher's Acknowledgments

Associate Editor: Elizabeth Stilwell

Development Editor: Rebecca Senninger

Technical Editor: Cecil Phillip

Proofreader: Debbye Butler

Production Editor: Mohammed Zafar Ali

Cover Image: © MandriaPix/Shutterstock

Leverage the power

Dummies is the global leader in the reference category and one of the most trusted and highly regarded brands in the world. No longer just focused on books, customers now have access to the dummies content they need in the format they want. Together we'll craft a solution that engages your customers, stands out from the competition, and helps you meet your goals.

Advertising & Sponsorships

Connect with an engaged audience on a powerful multimedia site, and position your message alongside expert how-to content. Dummies.com is a one-stop shop for free, online information and know-how curated by a team of experts.

- Targeted ads
- Video
- Email Marketing
- Microsites
- Sweepstakes sponsorship

20 MILLION PAGE VIEWS EVERY SINGLE MONTH

15 MILLION UNIQUE VISITORS PER MONTH

43% OF ALL VISITORS ACCESS THE SITE VIA THEIR MOBILE DEVICES

700,000 NEWSLETTER SUBSCRIPTIONS TO THE INBOXES OF *300,000* UNIQUE INDIVIDUALS EVERY WEEK